Introduction to Sociology
Group Activities

Lori Ann Fowler
Tarrant County College

WADSWORTH
CENGAGE Learning

Australia • Brazil • Japan • Korea • Mexico • Singapore • Spain • United Kingdom • United States

WADSWORTH
CENGAGE Learning

Introduction to Sociology: Group Activities
Lori Ann Fowler

For product information and technology assistance, contact us at
Cengage Learning Customer & Sales Support, 1-800-354-9706

For permission to use material from this text or product,
submit all requests online at **www.cengage.com/permissions**
Further permissions questions can be emailed to
permissionrequest@cengage.com

ISBN-13: 978-0-495-11556-4

ISBN-10: 0-495-11556-8

Wadsworth
10 Davis Drive
Belmont, CA 94002
USA

Cengage Learning is a leading provider of customized learning solutions with office locations around the globe, including Singapore, the United Kingdom, Australia, Mexico, Brazil, and Japan. Locate your local office at **international.cengage.com/region**

Cengage Learning products are represented in Canada by Nelson Education, Ltd.

To learn more about Wadsworth, visit **www.cengage.com/wadsworth**

Purchase any of our products at your local college store or at our preferred online store **www.ichapters.com**

Printed in the United States of America
3 4 5 6 7   14 13 12 11 10

ED309

Table of Contents

Chapter 1	Sociological Perspectives	1
Chapter 2	Research Methods	9
Chapter 3	Culture	19
Chapter 4	Socialization	26
Chapter 5	Society, Social Structure, and Interaction	36
Chapter 6	Groups and Organizations	44
Chapter 7	Deviance and Crime	51
Chapter 8	Global Stratification	59
Chapter 9	Social Class in the United States	66
Chapter 10	Race and Ethnicity	73
Chapter 11	Sex and Gender	80
Chapter 12	Aging and Inequality Based on Age	88
Chapter 13	The Economy and Work in Global Perspective	96
Chapter 14	Politics and Government in Global Perspective	109
Chapter 15	Families and Intimate Relationships	118
Chapter 16	Education	126
Chapter 17	Religion	135
Chapter 18	Health, Health Care, and Disability	142
Chapter 19	Population and Urbanization	150
Chapter 20	Collective Behavior, Social Movements, and Social Change	157
	Answer Key	164

Preface

This *Group Project Ancillary* is designed to help you review, study, and apply Introductory Sociology terms and concepts. After reading each chapter of your *Introductory Sociology Text*, refer to the corresponding chapter in this manual to help you review the material and asses how much you have retained and understand from your reading.

Each chapter in this ancillary includes the following sections:

A Video Clip Exercise – There are 1-2 exercises based on clips from *Wadsworth's Lecture Launchers* and *Wadsworth's Core Concepts* Video Collection Series for Introductory Sociology.

A Map the Stats Exercise – Using a current map and census data, students are asked to incorporate the data and discover findings and/or trends in the United States and the World.

A Case Study Exercise – Students are encouraged to visit current Internet Case studies and then answer questions using sociological concepts.

Group Quiz Exercises – Classes are encouraged to team up in groups to see who can answer the most correctly!

An Ethical Debate Exercise – Challenging topics are presented, and then students are divided into two groups and allowed to prepare for an in-class debate.

An Inside Class Activity – Students are asked to work on in-class group projects on various sociological topics.

An Outside Activity – Students are encouraged to perform research on various sociological topics outside of the traditional classroom setting.

Group Project Topics – Students are encouraged to work in groups on a research project. The topic and guidelines are provided, and their results are then presented to the class.

Ideas for Outside Reading – Sociological references are provided on specific topics when available.

I hope you find this guide helpful. I think all Sociology Instructors and Professors can find numerous uses for this guide inside the classroom setting. Please note that participating in the group quizzes alone will not prepare you fully for any sociological exam in your class.

Lori Fowler
Associate Professor, Sociology

CHAPTER ONE – SOCIOLOGICAL PERSPECTIVES
Video Clip Exercise

Watch the recommended video segments and then answer the questions below.

Video Segment 1:
Wadsworth's Lecture Launchers for Introductory Sociology:
"The Sociological Imagination"

Video Clip Exercise:
1. How does a parade serve as an opportunity to exercise one's sociological imagination?

2. Explain the relationship between social structure and parades.

3. Are there other social activities in which you can see a sociological example of our social structure?

Video Segment 2:
Wadsworth's Lecture Launchers for Introductory Sociology:
"The Feminist Perspective"

Video Clip Exercise:
1. What are some examples of gender differences in our society?

2. What are some ways in which gender affects our lives?

3. How important is gender socialization?

Map the Stat Exercise

SUICIDE DEATH RATE PER 100,000 (2002)

50 STATES

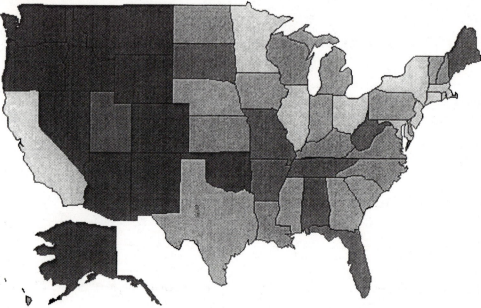

Suicide death rate per 100,000 (2002)		6.6 To	9.8 (10)
		9.9 To	11.2 (10)
		11.3 To	11.9 (10)
		12.7 To	14.0 (9)
		14.2 To	22.3 (11)

Map the Stats Exercise:

(1) According to the national map, suicide is least likely to occur in which states? Where is suicide most likely to occur?

(2) What environmental factors may lead to higher rates of suicide in the northwest region of the United States? What social factors may lead to the same? Why do you feel the suicide rate is lower in the northeastern region of the country?

Case Study Exercise

Visit these Internet Case studies and then answer the following questions.

Management: Is the IT Department Spying on You?

http://www.cioinsight.com/article2/0,1540,1957231,00.asp

http://www.govtech.net/magazine/channel_story.php?channel=4&id=94072

<u>Case Study Exercise:</u>

(1) Do you feel IT monitoring is an invasion of privacy or the necessary right of the employer? Explain your answer using article examples.

(2) Do you believe that the employer should notify employees if monitoring takes place? Should employees grant "permission" for this activity to take place, or is it an understood within company mainframes?

Group Quiz Exercise

Team up in groups to see who can answer the most correctly!

1. Which of the following statements BEST represents the Sociological Perspective?
    a) it focuses on relationships, not individuals
    b) in order to understand behavior, we must look at people in an ethnocentric way
    c) it views sociology as the science of the obscure
    d) it focuses on individuals, not relationships

2. The best definition of 'sociology' is which of the following?
    a) the subjective study of human society
    b) the objective study of human society
    c) the systematic study of human society
    d) the systematic and objective study of human society

3. In order to understand behavior, we must look at it in the context of _____ behavior?
    a) common
    b) unique
    c) group
    d) situational

4. Social scientists are expected to remain unbiased when conducting social research. This is the principle of_____?
    a) objectivity
    b) open reason
    c) verifiability
    d) idealism

5. Which type of reality is based on a direct result of our own experiences?
    a) altruistic
    b) experiential
    c) egoistic
    d) passivistic

6. Sociology is a unified discipline, meaning there are only a few theories a sociologist can choose from.
    a) True
    b) False

7. Women attempt suicide more than men, but men are more successful.
    a) True
    b) False

8. A Teenager is most likely to attempt suicide in the fall on a Tuesday.
    a) True
    b) False

9. Only 2 tablespoons of Pinesol or Draino can kill a small child?
   a) True
   b) False

10. The ultimate goal in sociology is to explain, predict, and understand behavior.
    a) True
    b) False

Ethical Debate Exercise

Divide the class into two groups and have each prepare for 15 minutes, allowing them to prepare for a class debate.

"Sociology is a social science that makes generalizations about people and their social life."

Group One:
Generalizations are a glorified method of stereotyping that is harmful to others.

Group Two:
Generalizations are NOT the same thing as stereotypes, therefore the practice is beneficial for society.

In Class Activity

As a group, select any newspaper or magazine article that depicts a recent criminal activity. Using your article, answer the following questions thoroughly.

1. How would a police officer view this particular incident? In other words, what *perspective* would the police officer reveal?

2. How would a psychiatrist view this particular article? In other words, what *perspective* would the psychiatrist reveal?

3. How would a sociologist view this particular article? In other words, what *perspective* would the sociologist reveal?

4. List all of the groups mentioned in the article. If only individuals are mentioned, what groups may they belong to?

5. Was this particular incident shaped by group membership or influence in any way?

6. What group membership may help to cure/prevent/solve any problems represented in this article?

Outside Activity

Research the topic "The Sociological Perspective" on the Internet using the following guidelines.

1. What is the Wikipedia (encyclopedia) definition of the Sociological Perspective?

2. What is the Wikipedia (encyclopedia) definition of the Sociological Imagination?

3. How are the two definitions similar or different? In many introductory sociology texts the two are used interchangeably. Should they be?

4. Do all sociologists view the world in the same way? Do an internet search to find your answer and site all examples.

Group Project Topics

Have each group member select one TV network channel. Watch the network for one hour 5 evenings in a row. Put together a coding sheet that will allow for the collection of basic data on all of the TV characters that are seen on that network.

Record the following for each TV characters on your network channel:

Age
Education
Ethnicity
Family Status
Family Size
Gender
Occupation
Race
Residential Pattern

Have each group member tabulate summary statistics and then compare the TV networks to determine which network is demonstrating the highest and the lowest social class, the greatest and the lowest gender ratios, and the greatest and least amount of racial representation.

Have the groups present their findings to the class.

Ideas for Outside Reading

Basirico, Laurence, A. 1993. "Sociology in Action: Activities for Students." *Harper Collins College Publishers*. New York, NY: Harper Collins.

DeVita, Philip, R. 1993. "Distant Mirrors: America as a Foreign Culture." *Wadsworth*. Belmont, Ca: Wadsworth, Inc.

Fowler, Lori, A. 2007. "Introduction to Sociology: A Real World Approach." *Wadsworth*. Belmont, Ca: Wadsworth, Inc.

Osborne, Richard. 2002. "Introducing Sociology." *Icon Books, Ltd.* UK, Europe: Macmillan Distribution Ltd.

Ruane, Janet, M. 1997. "Seeing Conventional Wisdom through the Sociological Eye." *Pine Forge Press*. Thousand Oaks, Ca: Sage.

CHAPTER TWO – RESEARCH METHODS
Video Clip Exercise

Watch the recommended video segments and then answer the questions below.

Video Segment 1:
Wadsworth's Lecture Launchers for Introductory Sociology:
"Elian Gonzalez: A Refugees Story"

Video Clip Exercise:
1. How is the conflict perspective illustrated in the case of Elian Gonzalez?

2. What are some examples of conflict in the Elian Gonzalez case?

3. How would you apply other sociological perspectives such as functionalism or interactionism to the Elian Gonzalez case?

Video Segment 2:
Wadsworth's Lecture Launchers for Introductory Sociology:
"Mixed Research Methods and New York City's Homeless"

Video Clip Exercise:
1. What are the research methods used by Kornblum and Williams?

2. Why would interviews be appropriate in this research effort?

3. What are the practical outcomes of this research?

Map the Stat Exercise

NUMBER OF PHYSICIANS PER 100,000 POPULATION

172 NATIONS OF THE WORLD

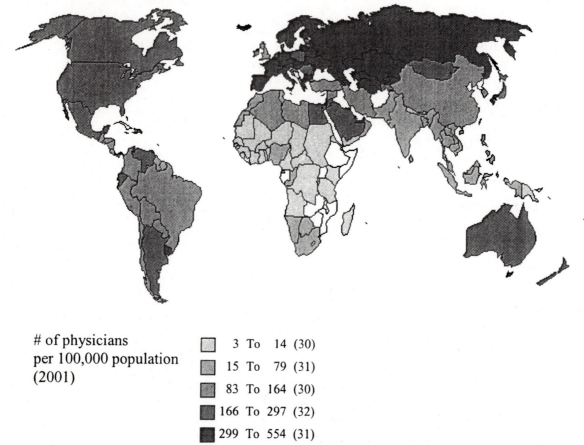

# of physicians per 100,000 population (2001)		
	3 To 14	(30)
	15 To 79	(31)
	83 To 164	(30)
	166 To 297	(32)
	299 To 554	(31)
	Missing	(18)

Map the Stats Exercise:

(1) Which countries in the world have the highest number of physicians per 100,000 population? Which countries have the last amount of physicians? Why is it important to compare the United States with countries like Canada and South America?

(2) Africa clearly has the last amount of physicians per 100,000 population. Why is this significant in these modern times? What would hinder doctors in Europe from moving into Africa to lend aid? Compare South America and Asia. What are the similarities?

Case Study Exercise

View this Internet Case study and then answer the following questions.

Stories told to us by people who have a friend or relative who has HIV.

http://www.positivesingles.com/AIDSstory

Case Study Exercise:

(1) After reading "My Father," do you feel this 9 year old girl should have been better prepared for her father's death? How old was she when she discovered he had died from the AIDS virus? As a sociological researcher, are you responsible for reporting this opinion, or keeping it to yourself?

(2) What happened in the case of "The Rape?" As a sociological investigator, are you responsible for reporting this type of information that is divulged to you? How?

Group Quiz Exercise

Team up in groups to see who can answer the most correctly!

1. The most widely used research method is the?
    a) experiment
    b) questionnaire
    c) participant observation
    d) existing sources

2. The _____ variable influences or causes changes to occur in the _____ variable.
    a) independent, dependent
    b) dependent, independent
    c) independent, associate
    d) associate, dependent

3. Which of the following study things that can be counted in numbers?
    a) qualitative analysis
    b) naturalistic analysis
    c) ethnomethodology
    d) quantitative analysis

4. Social scientists are expected to remain unbiased when conducting social research. This is the principle of_____?
    a) objectivity
    b) open reason
    c) verifiability
    d) idealism

5. Which of the following focuses on studying large scale organizations?
    a) macrolevel studies
    b) microlevel studies
    c) urban ecology
    d) none of the above

6. Which of the following determines the questions sociologists ask when doing research?
    a) the research perspective
    b) the scientific perspective
    c) the philosophical perspective
    d) the theoretical perspective

7. The _____ perspective studies at the microlevel:
    a) the conflict perspective
    b) the individual perspective
    c) the functionalist perspective
    d) the interactionist perspective

8. A _____ analysis would study items that cannot be counted in numbers.
   a) spurious
   b) unsymmetric
   c) qualitative
   d) quantitative

9. The key in doing experiments is:
   a) making sure the two groups are as different as possible
   b) making sure the two groups are as similar as possible
   c) making sure everyone is exposed to the cause
   d) none of the above

10. Identify the dependent variable in the following statement:
    Men are more likely than women to commit suicide.
    a) The dependent variable is gender.
    b) The dependent variable is suicide.
    c) The dependent variable is women.
    d) The independent variable is suicide.

Ethical Debate Exercise

Divide the class into two groups and have each prepare for 15 minutes, allowing them to prepare for a class debate.

The Hawthorne effect - an increase in worker productivity produced by the psychological stimulus of being singled out and made to feel important.
The Hawthorne Effect was first noticed in the Hawthorne plant of Western Electric.
Production increased not as a consequence of actual changes in working conditions introduced by the plant's management but because management demonstrated interest in such improvements.

Group One:
The Hawthorne Effect is a real phenomenon in which subjects change their behavior when they know they are being observed. Therefore, research involving video cameras is too biased and should NOT be conducted.

Group Two:
The Hawthorne Effect fades over time, and should NOT sway researchers from using video cameras in field research.

In Class Activity

As a group, read over the 8 steps of conducting research discussed in class. Choose a topic that you as a group wish to apply the 8 steps to.

8 Steps in Conducting Research

8 Steps
1- Defining the Topic
2- Reviewing the Literature
3- Identifying Concepts and Variables
4- Forming hypotheses
5- Choosing a Research Design
6- Collecting Data
7- Analyzing the Data
8- Drawing Conclusions

Answer the following questions:
1. As a team of researchers, your first step involves choosing a topic of study.
a) Which topic have you chosen? _____
b) Why is it worth studying? _____
c) Why is your topic timely and important? _____

2. **Imagine** you were visiting your local library in order to research your topic. Write down ten findings you may discover regarding your topic. For example, depression leads to suicide among teens.

1-	6-
2-	7-
3-	8-
4-	9-
5-	10-

3. After researching your topic, you must now identify your independent and dependent variables.
*Remember you are stating that the independent variable is causing the change in the dependent variable.

Example: Depression
            Anxiety
            Poverty   ----------------------------→ Suicide
            Drugs
            Alcohol

Now you choose 5 independent variables from your #2 results that will lead to your topic.
     Findings from Lit Review   Topic

            _____
            _____
            _____   ------------------------→ _____
            _____
            _____

4. Once you have identified your independent and dependent variables, you must translate them into hypotheses. Write down the above 5 variables in bold, statement form.

Example: Depression among women causes suicide.

_____
_____
_____
_____
_____

5. After completing the above steps, you must decide on a plan for collecting data.
Who specifically are you going to study? _____
Why them? _____

6. Which research method are you going to employ? You can use only one or several. Why?
Paper and pencil questionnaire?

Interview?
Field Research?
Experiment?

7. Once you have gathered your data, you can prove your hypotheses true or false. Which do you *really* think will occur with each of your 5 hypotheses?

If they are true, you accept.
If they are false, you reject.

_____   accept or reject?
_____   accept or reject?
_____   accept or reject?
_____   accept or reject?
_____   accept or reject?

8. Finally, you must draw conclusions.
    a) How well do you feel your findings would represent the total population? Why or why not?

    b) What would you do differently next time?

Outside Activity

You need to go out onto the campus ALONE: into the student center, library, bookstore, cafeteria, faculty building, or outdoors and observe smoking, flirting, lying, fighting, cheating, sharing, whatever!

Use this worksheet to organize the data in your field exercise, recording the following information, as carefully as possible.

You need to observe for 35 minutes! You must record every minute! See just how interesting, difficult, and fun observing behavior can be. REMEMBER: you do not want them to know they are being observed!

EVERY MINUTE MUST BE RECORDED IN FIELD NOTES!

For Example:

9:01 saw woman moving towards door

9:02 woman looked at man

9:03 man glanced back at woman

Record every minute of your 35 minutes here just like above example:

___ ___

___ ___

___ ___

___ ___

___ _____

___ _____

___ _____

___ _____

___ _____

___ _____

___ _____

___ _____

___ _____

___ _____

___ _____

___ _____

___ _____

___ _____

___ _____

___ _____

___ _____

___ _____

___ _____

___ _____

___ _____

___ _____

___ _____

———
———
———
———
———
———
———
———

Date:

Behavior expected to see:

Field Location:

Time begun:

Time ended:

Number of people observed:

Weather:

Initial thoughts and feelings as you begin the observation:

Spatial arrangement of location (create a sketch below):

```
┌─────────────────────────────────────────────────┐
│                                                   │
│                                                   │
│                                                   │
│                                                   │
│                                                   │
│                                                   │
└─────────────────────────────────────────────────┘
```

Overall emotional tone of the environment:

Nonverbal behavior noted:

Verbal interaction noted:

Any observable deviance?

Group Project Topics

Have each group select one research topic of their choice. Collect newspaper articles, empirical articles, and photos pertaining to the topic. Put together a PowerPoint presentation demonstrating the eight steps involved in research and how the collection of data would take place. Be sure to identify independent and dependent variables.

Be sure to follow and demonstrate these eight steps in your presentation:

1- Defining the Topic
2- Reviewing the Literature
3- Identifying Concepts and Variables
4- Forming hypotheses
5- Choosing a Research Design
6- Collecting Data
7- Analyzing the Data
8- Drawing Conclusions

Have the groups present their research to the class.

Ideas for Outside Reading

Basirico, Laurence, A. 1993. "Sociology in Action: Activities for Students." *Harper Collins College Publishers*. New York, NY: Harper Collins.

Eichler, Margrit. 1988. "Nonsexist Research Methods." Winchester, Mass.: Unwin Hyman.

Fowler, Lori, A. 2007. "Introduction to Sociology: A Real World Approach." *Wadsworth*. Belmont, Ca: Wadsworth, Inc.

Roethlisberger,F.J. & Dickson,W.J. 1939. "Management and the Worker." Cambridge, Mass.: Harvard University Press.

Whyte, William Foote. 1981. "Street Corner Society." Chicago: University of Chicago Press.

Zimbardo, Philip, G. "Pathology of Imprisonment." *Society*, Vol. 9 (April 1972):4-8.

CHAPTER THREE - CULTURE
Video Clip Exercise

Watch the recommended video segments and then answer the questions below.

Video Segment 1:
Wadsworth's Lecture Launchers for Introductory Sociology:
"Culture Shock"

Video Clip Exercise:
1. How does culture shock affect an individual?

2. Why is the expression "a fish our of water" descriptive of culture shock?

3. Give some examples of culture shock.

Video Segment 2:
Wadsworth's Sociology: Core Concepts:
"Culture"

Video Clip Exercise:
1. What are some examples of norms in our everyday life? Define culture.

2. According to Kornblum, what are the major dimensions of culture?

3. How is material culture different from nonmaterial culture?

Map the Stat Exercise

% FEMALE HEADED HOUSEHOLDS BELOW THE POVERTY LINE

50 STATES

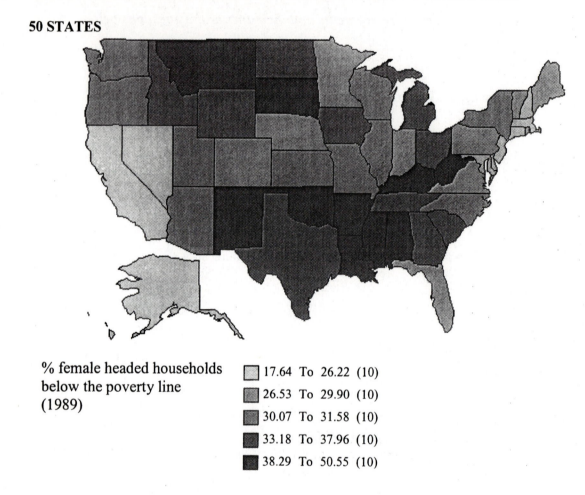

% female headed households
below the poverty line
(1989)

17.64 To 26.22 (10)
26.53 To 29.90 (10)
30.07 To 31.58 (10)
33.18 To 37.96 (10)
38.29 To 50.55 (10)

Map the Stats Exercise:

(1) The majority of female headed households living below the poverty line are located in which region of the country? Which two states seem to be surrounded by higher levels of poverty stricken households?

(2) Why do you feel the southern region of the United States is so inundated with poverty? Could there be other influences at work here? What are your thoughts?

Case Study Exercise

View this Internet Case study and then answer the following questions.

Material Culture Narratives

https://tspace.library.utoronto.ca/citd/holtorf/6.8.html

Case Study Exercise:

(1) What can be used as reference points for messages about continuity or change of social identity?

(2) Communities or individuals who placed secondary burials in the mounds of megaliths, too, may have used the ancient mounds in order to tell a _____, either about themselves or about the buried individuals. Mounds alone can function as mnemonics of _____.

(3) Affirming apparently ancient traditions can be a powerful method to _____ change.

Group Quiz Exercise

Team up in groups to see who can answer the most correctly!

1. Which of the following terms represents a type of prejudice that says my culture's ways are right and other's ways are wrong?
    a. Cultural relativism
    b. Ethnocentrism
    c. Dominant culturalism
    d. Culture discrimination theory

2. Which of the following statements is <u>false</u>?
    a. We are born with elements of culture
    b. We must learn our culture after birth
    c. Culture is the total lifestyle of a people
    d. Each culture is shaped by physical and social factors

3. Which of the following represents anything made with human hands?
    a. Non–material culture
    b. Norms
    c. Material culture
    d. Tangible establishments

4. Which of the following type of culture is opposed to the dominant culture?
    a. A Counterculture
    b. A Subculture
    c. Competitive cultures
    d. Quasi–dominant cultures

5. Which of the following terms represents judging each culture from its own viewpoint without imposing judgment?
    a. Ethnocentrism
    b. Cultural relativism
    c. Normalcy
    d. Fragmentation

6. Which of the following examples best describes a <u>value</u> in American society?
    a. "Ghosts do not exist."
    b. "Success comes from hard work."
    c. "Keep your elbows off the table."
    d. "Stop at a red light."
    e. "Molestation is wrong."

7. Which of the following examples best describes a <u>folkway</u> in American society?
    a. "Ghosts do not exist."
    b. "Success comes from hard work."
    c. "Keep your elbows off the table."
    d. "Stop at a red light."
    e. "Molestation is wrong."

Ethical Debate Exercise

Divide the class into two groups and have each prepare for 15 minutes, allowing them to prepare for a class debate.

The World Health Organization (WHO, 1997) defined female genital mutilation (FGM) as all procedures involving partial or total removal of the external female genitalia or other injury to the female genital organs whether for cultural or other non-therapeutic reasons.

Group One:
FGM for women in Africa is NOT different than male circumcision in the United States.

Group Two:
FGM for women in Africa IS different than male circumcision in the United States.

In Class Activity

You are a very influential character within this generation and you are going to create an event that may change our society! You have decided to plan an event unlike any other. YOU CANNOT plan a wedding, birthday party, funeral, or anything we are already familiar with in America today. Plan a new event of your very own! Celebrate or mourn something unique- whatever you want – however you want!

1) You cannot memorialize **ANYTHING** anyone has done before. You must create this celebration or memorial from scratch. What are you going to do?

2) How are you going to get people to attend? **YOU CANNOT SEND OUT INVITATIONS or FLIARS!** (That has been done for ages)! Who are you going to invite? Why them? How are you going to get them there?

3) What are you going to wear? Why? What will the guests wear? Why? Make **EACH PIECE** of clothing have some significance. (For example; I would wear red because_____).

4) Would you exchange gifts, money, or any other items at this event? If so, why that or those? What do you plan to eat at this event? Why that?

5) Do you think this ceremony is one that should be passed on to future generations? Why or why not? How would you teach others to carry on the tradition?

6) Look over all of your answers above. Place each item you have listed above in the appropriate columns below.

MATERIAL CULTURE   **NON-MATERIAL CULTURE**

Group Project Topics

Planning a Culture Feast

Now you are responsible for bringing to life in the classroom what you planned above! You are going to demonstrate your celebration or mourning in class during a culture feast.

You should plan to have each group member prepare the following items:

Creative Idea
Telling the Class why you should have such an event
Invitation style/demonstration
Clothing Style Demonstrated
Food Demonstrated/Shared with Class
Example of your Exchange Item shown in class

**Make sure you DEMONSTRATE each of the items above.
You are NOT responsible for feeding the entire class! Bring enough to demonstrate as an example; you'll be amazed at how much food you have! Remember serving spoons and plates. Each person should bring their own drink. Each person should dress appropriately.**

Ideas for Outside Reading

Basirico, Laurence, A. 1993. "Sociology in Action: Activities for Students." *Harper Collins College Publishers*. New York, NY: Harper Collins.

Brym, Robert J. and Lie, John. 2007. "Sociology." Thomson Wadsworth. Belmont, CA: Wadsworth.

DeVita, Philip, R. 1993. "Distant Mirrors: America as a Foreign Culture." *Wadsworth*. Belmont, Ca: Wadsworth, Inc.

Fowler, Lori, A. 2007. "Introduction to Sociology: A Real World Approach." *Wadsworth*. Belmont, Ca: Wadsworth, Inc.

CHAPTER FOUR – SOCIALIZATION
Video Clip Exercise

Watch the recommended video segments and then answer the questions below.

Video Segment 1:
Wadsworth's Lecture Launchers for Introductory Sociology:
"The Nature vs. Nurture Debate"

Video Clip Exercise:
1. Who was an early proponent of nature's influence?

2. Who were early proponents of the nurture side of the nature vs. nurture debate?

3. Which do you feel is more influential in development, nature or nurture? Why?

Video Segment 2:
Wadsworth's Sociology: Core Concepts:
"Socialization"

Video Clip Exercise:
1. What happens to children with little or no social contacts?

2. What are the agencies of socialization? Which agency is considered the primary?

3. What is the purpose of schools in the socialization process?

Map the Stat Exercise

PERCENT WHO THINK IT VERY IMPORTANT THAT A CHILD BE OBEDIENT

172 NATIONS OF THE WORLD

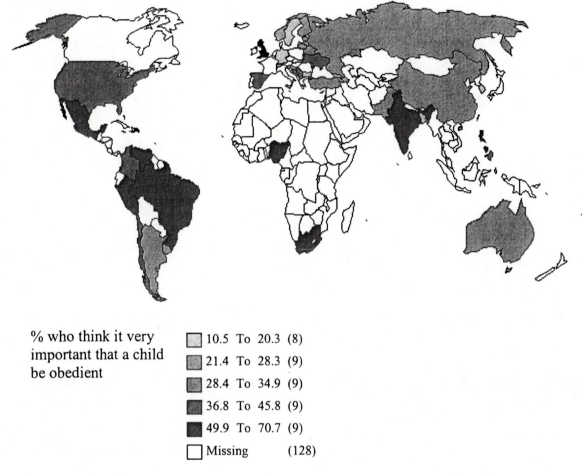

% who think it very
important that a child
be obedient

- ☐ 10.5 To 20.3 (8)
- ▨ 21.4 To 28.3 (9)
- ▨ 28.4 To 34.9 (9)
- ▩ 36.8 To 45.8 (9)
- ■ 49.9 To 70.7 (9)
- ☐ Missing (128)

Map the Stats Exercise:

(1) What are your thoughts surrounding the word "obedient?" Do you feel this question is a fair question to be asking on an international level? Why or why not?

(2) Which two large countries are least likely to think that is important for children to be obedient? Do you think this means these countries wish for their children to be lawless? What percent think it very important that a child be obedient in the United States?

Case Study Exercise

Visit these Internet Case studies and then answer the following questions.

Kids & Viewing TV Violence

<u>http://www.washingtonpost.com/wp-dyn/content/discussion/2006/04/07/DI2006040701360.html</u>

<u>http://www.washingtonpost.com/wp-dyn/content/article/2006/04/10/AR2006041001308.html</u>

<u>Case Study Exercise:</u>

(1) The _____ television shows children watch and _____ can be just as important as the amount of time they spend in front of the tube.

(2) What is David Bickham's main job as a researcher?

(3) Children who watch _____television programs -- especially those who watch such shows alone -- spend _____with friends than children who watch a lot of nonviolent programs.

(4) How much time does the average child spend watching television?

(5) Do you feel the media has more of an impact on one's development than other agents of socialization? Use support from this article and the text to defend your answer.

Group Quiz Exercise

Team up in groups to see who can answer the most correctly!

1. If someone were to ask you which children were most susceptible to TV violence, which of the following groups would you state were the most affected?
    a. Girls ages 6–8
    b. Girls ages 3–5
    c. Boys ages 3–5
    d. Boys ages 6–8

2. Which of the following terms represents ways in which a teacher evaluates non–academic behaviors?
    a. Norm–evaluation
    b. Ethnomethodology
    c. Hidden curriculum
    d. Valuatory judgment

3. What does socialization do?
    a. It controls us
    b. It provides a set of socially accepted pattern of thinking and acting
    c. It allows for culture to be passed on
    d. All of the above

4. Who conducted the BoBo Doll experiment, demonstrating that children model behavior they see on television?
    a. Emile Durkheim
    b. Karl Marx
    c. Albert Bandura
    d. Susan Brownmiller

5. Which of the following harmful effects can occur from sustained exposure to media violence?
    a. learning and imitation
    b. desensitization
    c. fear
    d. all of these choices

6. Entertaining TV shows that contain violence do not harm young people.
    a. True
    b. False

7. The National Institute of Mental Health discovered that violence on television leads to aggressive behavior. This aggression also occurs in adults.
    a. True.
    b. False.

8. The level of violence during Saturday morning cartoons is higher than the level of violence during Prime Time.
   a. True
   b. False

9. As a result of media violence, children learn that there are few if any repercussions for committing violent acts.
   a. True
   b. False

10. Which of the following would describe the process by which people develop personal identities and learn the ways of a particular group or society?
   a. Control
   b. Socialization
   c. Authority
   d. Transference

Ethical Debate Exercise

Divide the class into two groups and have each prepare for 15 minutes, allowing them to prepare for a class debate.

There has been much controversy among psychologists and sociologists in the late 20th century concerning whether some people are genetically disposed to crime or whether illegal acts have their origin in one's upbringing and environment.

Group One:
Individual's who commit criminal acts do so because they are raised in an environment that is conducive to such behavior.

Group Two:
Individual's who commit crime do so because they are biologically or genetically different than non-deviant individuals.

In Class Activity

In groups, answer the following questions about socialization.

1. Which of the following would you claim has the MOST influence on a person's behavior? You may only choose one!

Family　　　　　　**Schools**　　　　　　**Peers**　　　　　　**Mass Media**

Why? Explain:

2. How do you feel a person learns how to read and understand body language?

3. How does a person learn to act violently? Does TV really play a role? Are some people just born violent, in your opinion?

4. Choose a magazine at random. Find (3) three articles demonstrating socialization at work in order to complete the following. Tear out each article and turn them in with this sheet.

ARTICLE ONE
  　　　a) **Name of article:**
  　　　b) **Summary of article:**

  　　　c) **Is the article demonstrating positive or negative socialization?**

ARTICLE TWO

      a) Name of article:

      b) Summary of article:

      c) Is the article demonstrating positive or negative socialization?

ARTICLE THREE

      a) Name of article:

      b) Summary of article:

      c) Is the article demonstrating positive or negative socialization?

Outside Activity

Serial Killer Exercise

Richard Ramirez
John Wayne Gacy
Albert Fish
Genene Jones
Jack the Ripper
David Berkowitz
Elizabeth Bathory

(1) Choose one killer from the list above to conduct your research. How EXACTLY did the murders happen? Be sure to find out the age of the killer at the time of his or her first murder.

(2) What seemed to be the motive involved in the murders? Is there more than one?

(3) Describe each of the victims:
    a. What did they have in common?

    b. What was different about each of them?

    c. Anything interesting?

(4) What was your killers preferred style of death? Why is this important?

(5) What were your killers displayed post offense behaviors? i.e. What did he or she do after the murders?

(6) Were alcohol or drugs involved before, during, or after any or all of the murders?

(7) Do you feel your murderer was made or born a murderer? EXPLAIN!

Group Project Topics

The Real Message Behind the Music

Listen to 20 seconds of random music. I prefer you select 10 different genres. Listen carefully, without singing along. Write down the REAL MESSAGE contained within the lyrics.

Song Title:
Real Message:

Song Title:
Real Message:

Song Title:
Real Message:

Song Title:
Real Message:

Song Title:
Real Message:

Song Title:
Real Message:

Song Title:
Real Message:

Song Title:
Real Message:

Song Title:
Real Message:

Song Title:
Real Message:

Ideas for Outside Reading

Basirico, Laurence, A. 1993. "Sociology in Action: Activities for Students." *Harper Collins College Publishers*. New York, NY: Harper Collins.

Brym, Robert J. and Lie, John. 2007. "Sociology." Thomson Wadsworth. Belmont, CA: Wadsworth.

Fowler, Lori, A. 2007. "Introduction to Sociology: A Real World Approach." *Wadsworth*. Belmont, Ca: Wadsworth, Inc.

Osborne, Richard. 2002. "Introducing Sociology." *Icon Books, Ltd*. UK, Europe: Macmillan Distribution Ltd.

Ruane, Janet, M. 1997. "Seeing Conventional Wisdom through the Sociological Eye." *Pine Forge Press*. Thousand Oaks, Ca: Sage.

CHAPTER FIVE – SOCIETY, SOCIAL STRUCTURE, AND INTERACTION
Video Clip Exercise

Watch the recommended video segments and then answer the questions below.

Video Segment 1:
Wadsworth's Lecture Launchers for Introductory Sociology:
"Social Interaction: The Ropes Course"

Video Clip Exercise:
1. How does social structure relate to function in the ropes course?

2. How is leadership determined in the ropes course?

3. What is the role of socialization in shaping our lives?

Video Segment 2:
Wadsworth's Sociology: Core Concepts:
"Social Interaction"

Video Clip Exercise:
1. Name the four principles of interaction.

2. What is the relationship between the pleasure principle and human interaction?

principle of interaction deals with cost-benefit analysis?

Case Study Exercise

View this Internet Case study and then answer the following questions.

Lies: Are We All Masters of Deception?

http://www.kevinhogan.com/lies.htm

<u>Case Study Exercise:</u>

(1) Why does the research state that people lie chronically? Are they mentally ill?

(2) Psychologists have long known that some deception is a normal, healthy part of human behavior, often starting in children as young as _____. In adulthood, most people lie routinely, if usually harmlessly, throughout the day.

(3) What is the average lie rate among men and women in daily conversation?

(4) Why do people lie?

(5) What do chronic liars demonstrate on psychological tests?

Group Quiz Exercise

Team up in groups to see who can answer the most correctly!

1. Women smile to show?
    a. Their teeth
    b. Respect
    c. Submission
    d. Flirtation

2. In discussing the use of space, using more space conveys a message of _____?
    a. Femininity
    b. Personal importance
    c. Gender accomplishment
    d. Submission

3. When discussing eye contact, we mentioned that people look at one another during conversations about 30–60% of the time. If you look at someone more than 60% of the time, you want _____?
    a. To fight
    b. Their body
    c. To lie
    d. To mimic their behavior

4. The toast originated long ago. Why?
    a. To ensure the drinks sloshed together, mixing any poisons.
    b. To ensure that rival enemies became friends.
    c. To elevate the elbow, and allow others to see if weapons were stowed below, under the armpit.
    d. All of the above.

5. Which of the following statements are true of social interaction?
    a. A group is made up of the three or more in sociology.
    b. You have to be physically close to someone in order to interact with him or her.
    c. Being near others always means social interaction will occur.
    d. None of the above.

6. Which of the following terms represents viewing social interaction in terms of theatrical performance?
    a. Defining the situation
    b. Dramaturgical analysis
    c. Symbolic interaction
    d. Impression management

7. We discussed the meaning and power of colors in class. McDonald's and Doctor's offices are painted certain colors for certain reasons. Which of the following statements are correct?
    a. McDonald's is painted red because red stimulates, excites, and acts as an irritant.
    b. Doctor's offices are painted white because this color is strengthening and purifying.
    c. Both a and b.
    d. None of these choices.

8. Which of the following refers to the science of body language?
    a. Rogue analysis
    b. Kinesics
    c. Sociology
    d. Graphoanalysis

9. Which of the following traits may be defined as rogue?
    a. Nose, eyebrow, and tongue piercing
    b. Pink Mohawks
    c. Very tall men
    d. All of these choices

10. Proxemics describes our use of space.
    a. True
    b. False

Ethical Debate Exercise

Divide the class into two groups and have each prepare for 15 minutes, allowing them to prepare for a class debate.

Exchange theorists argue that ALL social relationships involve give and take. If you give to others, you expect something in return. With payoffs, relationships will endure. Without them, they will not.

Group One:
All human relationships are based on exchange theory, and when the relationship becomes un-balanced, it will come to an end.

Group Two:
Not all relationships are based on exchange, and some acts can be describe un-selfish.

In Class Activity

Your group needs to select one magazine at random. You must start with the cover and move page by page, without skipping a single advertisement. Your group members need to decide which American value the magazine is trying to demonstrate on each of the 25 pages you go through. For example, if the magazine page is selling perfume, perhaps all members will agree the American value is "beauty or the fear of odor."

Values: general agreements about what is good or bad, right or wrong.

Magazine Page 1   Value Demonstrated:
Magazine Page 2   Value Demonstrated:
Magazine Page 3   Value Demonstrated:
Magazine Page 4   Value Demonstrated:
Magazine Page 5   Value Demonstrated:
Magazine Page 6   Value Demonstrated:
Magazine Page 7   Value Demonstrated:
Magazine Page 8   Value Demonstrated:
Magazine Page 9   Value Demonstrated:
Magazine Page 10   Value Demonstrated:
Magazine Page 11   Value Demonstrated:
Magazine Page 12   Value Demonstrated:
Magazine Page 13   Value Demonstrated:
Magazine Page 14   Value Demonstrated:
Magazine Page 15   Value Demonstrated:
Magazine Page 16   Value Demonstrated:
Magazine Page 17   Value Demonstrated:
Magazine Page 18   Value Demonstrated:
Magazine Page 19   Value Demonstrated:
Magazine Page 20   Value Demonstrated:
Magazine Page 21   Value Demonstrated:
Magazine Page 22   Value Demonstrated:
Magazine Page 23   Value Demonstrated:
Magazine Page 24   Value Demonstrated:
Magazine Page 25   Value Demonstrated:

Outside Activity

Observe people on campus walking, sitting, talking, or smoking. What can you tell about their behavior? What impression are they trying to create in the minds of others?

Watch 5 individuals on campus for 5 minutes each.
Record information for the following:

<u>Walking Style</u>   <u>Body Language</u>   <u>Gestures</u>   <u>Other</u>

Person #1

Person #2

Person #3

Person #4

Person #5

Group Project Topics

Breaking Norms

Now you are responsible for breaking norms outside of the classroom.

You should plan to have each group member break at least three norms:

Pay attention to the reactions of those around you. Each of you had to design 3 norms that you would break in public. Write down the 3 norms you each decided to break.

(1) _____   (1) _____   (1) _____
(2) _____   (2) _____   (2) _____
(3) _____   (3) _____   (3) _____

Now, discuss the reactions of others as you broke these norms. Were there any unexpected reactions? How did you feel while breaking these norms?

Person 1)

_____
_____
_____

Person 2)

_____
_____
_____

Person 3)

_____
_____
_____

Ideas for Outside Reading

Brym, Robert J. and Lie, John. 2007. "Sociology." Thomson Wadsworth. Belmont, CA: Wadsworth.

Fowler, Lori, A. 2007. "Introduction to Sociology: A Real World Approach." *Wadsworth.* Belmont, Ca: Wadsworth, Inc.

Hall, Edward, T. 1969. "The Hidden Dimension." Anchor Books. Garden City, New York. Doubleday and Co.

Scheflen, Albert, E. 1972. "Body Language and the Social Order." Englewood Cliffs, New Jersey. Prentice Hall.

Solomon, Jack. 1990. "The Signs of Our Time. The Secret Meanings of Everyday Life." New York, N.Y. Harper and Row Publishers.

CHAPTER SIX – GROUPS AND ORGANIZATIONS
Video Clip Exercise

Watch the recommended video segments and then answer the questions below.

Video Segment 1:
Wadsworth's Lecture Launchers for Introductory Sociology:
"Bystander Apathy"

Video Clip Exercise:
1. What is bystander apathy?

2. What explanation does the Kitty Genovese example offer as to why groups of people are apathetic?

3. In the absence of clearly defined responsibility, what happens?

Video Segment 2:
Wadsworth's Lecture Launchers for Introductory Sociology:
"The McDonaldization of Society"

Video Clip Exercise:
1. What is McDonaldization?

2. What are examples of McDonaldization illustrated in this video segment?

3. Where else in our culture do you see McDonaldization occurring?

Case Study Exercise

Visit these Internet Case studies and then answer the following questions.

McDonald's, McDonaldland, and McDonaldization

http://www.peaceaware.com/McD/

Case Study Exercise:

(1) This author claims that Ronald McDonald is not popular everywhere around the globe. What can you give as an explanation for this? Does this mean that McDonaldization is not a real phenomenon?

(2) What is your opinion of this researcher's modifications of McDonald's characters? Do you feel his modifications are realistic?

Group Quiz Exercise

Team up in groups to see who can answer the most correctly!

1. Individuals who share similar characteristics but do not interact would be an example of:
   a) aggregate
   b) category
   c) group
   d) assemblage

2. People who temporarily share the same physical space but have no sense of belonging together are a(n):
   a) aggregate
   b) category
   c) group
   d) assemblage

3. As a college student in the U.S., you are part of:
   a) a category
   b) an aggregate
   c) a group
   d) an assemblage

4. Intimacy and face-to-face interaction are characteristics of a(n):
   a) aggregate
   b) primary group
   c) secondary group
   d) category

5. Secondary groups:
   a) are characterized by intimate relationships and cooperation
   b) are formal and impersonal
   c) provide members with views and values to internalize
   d) often form out of primary groups

6. All except one of the following are examples of secondary groups. Which one should not be included?
   a) a college classroom
   b) the Democratic Party
   c) a friendship group
   d) the Republican Party

7. Although secondary groups are necessary, they often:
   a) do not satisfy our need for intimate associations
   b) lead to oligarchy
   c) become very informal and personal
   d) require too much loyalty

8. Voluntary associations:
    a) are unusual in the United States
    b) always have professional staff
    c) are organized on the basis of some mutual interest
    d) are usually political in nature

9. The leaders of voluntary associations:
    a) are usually chosen for their abilities
    b) have difficulty maintaining their position at the top
    c) are likely to grow distant from their members
    d) are often undermined

10. _____ refers to how organizations come to be dominated by a small, self-perpetuating elite.
    a) the iron law of oligarchy
    b) the Peter Principle
    c) the Thomas Theorem
    d) the power elite

Ethical Debate Exercise

Divide the class into two groups and have each prepare for 15 minutes, allowing them to prepare for a class debate.

Alienation: Man is alienated from the object he produces, from the process of production, from himself, and from the community of his fellows.

We, as a people, have separated ourselves so much from real life that it has become quite easy to believe that this life is just a game that someone else is playing.

Group One:
We are more alienated now than ever before. Our current technologies have created this problem.

Group Two:
We are less alienated than ever before. Our current technologies allow us to be one another than ever before.

In Class Activity

As a group, read the facts below and then answer the following questions thoroughly.

British authors studied the death statistics of the 19,721 people recorded as having committed suicide between 1988 and 1992. They concluded that 62 pairs of people had planned to kill themselves together. Forty-eight pairs were husband and wife, five were close blood relatives, four were lovers, three were cohabiting couples, and two were friends. Fifty-seven pairs lived in the same household, and only two people lived alone, in contrast to lone suicides, of which only half are married and a quarter live alone.

The ratio of males to females was one to one, another difference from lone suicides, where it is 3 to one in favor of males. The average age of people committing suicide together was higher, and nearly half the sample was in social class 1 or 2. Eleven people worked in professions relating to medicine.

Couples used non-violent methods, which permitted them to die painlessly and simultaneously. Suicide notes were left in 52 pacts, and both partners signed in 33 of these. In all, 69 percent signed a suicide note, again a higher proportion than among lone suicides (30-40 percent).

Fifteen of the married couples were remembered by witnesses as talking of "dying together" and "not bearing to be parted". This suggests that there was no coercion, say the authors.

1. Who are the most likely to engage in a suicide pact?

2. What is the main motivation for engaging in a suicide pact? Why would married couples engage in such behavior?

3. If Durkheim's theory regarding Suicide states that the more people feel connected the less likely they are to commit suicide, does it still apply to suicide pacts? Why or why not?

4. What is a possible solution to pact suicides?

Outside Activity

Research the topic "Corporate Culture" on the Internet using the following guidelines.

1. What is the definition of Corporate Culture? How is corporate culture expressed?

2. Corporate culture should not be confused with corporate mission. What is the difference?

3. How does a corporation operate on both conscious and unconscious levels?

Group Project Topics

Have each group member select one aspect of bureaucracies to do research. Put together a poster display for the entire class displaying each of the five characteristics below.

Display each of these characteristic in poster format:

Clear Cut Levels arranged in Hierarchical Order
A Division of Labor
Written Rules
Written Records
Impersonality

Have each group member summarize their findings and then compare
share them with the rest of the class.

Ideas for Outside Reading

Durkheim, Emile. 1979. *Suicide. A Study in Sociology.* New York: Free Press.

Engels, Frederick. 1986. *The Origin of the Family, Private Property and the State.* New York: Penguin Books.

Weber, Max. 2002. *The Protestant Ethic and the Spirit of Capitalism and Other Writings.* New York: Penguin Books.

Frank, Thomas. 2005. *What's the Matter With Kansas?* New York: Henry Holt and Company.

Collins, Randall and Michael Makowsky. 2005. The Discovery of Society. 7th Edition. University of Pennsylvania Press.

CHAPTER SEVEN – DEVIANCE AND CRIME
Video Clip Exercise

Watch the recommended video segments and then answer the questions below.

Video Segment 1:
Wadsworth's Lecture Launchers for Introductory Sociology:
"The Criminal Justice System: Plea Bargaining"

Video Clip Exercise:
1. What are the functions of plea bargaining within the criminal justice system?

2. What are the dysfunctions of plea bargaining for the larger society?

3. What is the third time offender law?

Map the Stat Exercise

PERCENT WHO BELIEVE EUTHANASIA IS OK

172 NATIONS OF THE WORLD

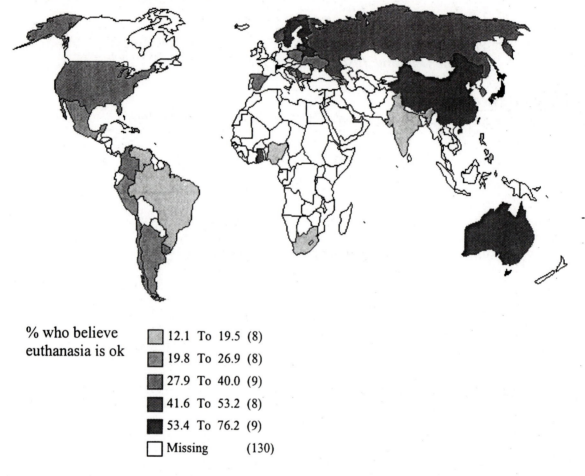

% who believe euthanasia is ok

- 12.1 To 19.5 (8)
- 19.8 To 26.9 (8)
- 27.9 To 40.0 (9)
- 41.6 To 53.2 (8)
- 53.4 To 76.2 (9)
- Missing (130)

Map the Stat Questions:

(1) Which nation has the highest acceptance of euthanasia? Which nations have the least acceptance of euthanasia? What are your feelings toward the practice of euthanasia? Why?

(2) What kinds of pressure do increasing older populations put on government services and allocation of funds?

Case Study Exercise

Visit this Internet Case study and then answer the following questions.

Why Is Killing for Capital
Not a Capital Crime?

http://www.reclaimdemocracy.org/corporate_accountability/ford_firestone_manslaughter.
html

Case Study Exercise:

(1) If the Ford Executives knew they had a dangerous car, should they have been found guilty of murder, even though the deaths were the results of collisions?

(2) Is it fair to blame a single CEO or designer for the activities of an entire company?

(3) Who should have cured this problem, and how?

Group Quiz Exercise

Team up in groups to see who can answer the most correctly!

1. In order for an act to be defined as a crime, it must be seen as _____ and
_____?
   a) corrupt & harmful
   b) pathetic and evil
   c) intentional and inexcusable
   d) discriminatory and devastating

2. According to the Crime Clock, one crime index offense occurs every _____?
   a) 2 minutes
   b) 2.7 seconds
   c) 4 minutes
   d) 4 seconds

3. In our discussion of homicide, you are much more likely to be killed by a _____ than a
_____?
   a) knife, gun
   b) husband, wife
   c) relative, stranger
   d) plane crash, auto crash

4. In our discussion of punishment, we discussed several things. Which among the following are
true?
   a) A person is likely to serve one–third of their sentence today
   b) If one is sentenced to two 99 year sentences, concurrently, they are likely to serve 66
   years
   c) If one is sentenced to two 99 year sentences, consecutively, they are likely to serve
   33 years
   d) All of the above are true

5. Which of the following typically involve more than one offender?
   a) homicide
   b) rape
   c) robbery
   d) assault

6. Which of the following are more likely to commit embezzlement and fraud?
   a) executives
   b) women
   c) men
   d) teenagers

7. Which type of crime operates for profit or power that seeks to obtain immunity from the law through fear and corruption?
>   a) organized crime
>   b) white collar crime
>   c) pink collar crime
>   d) corporate crime

8. Which of the following is the most costly in dollars?
>   a) organized crime
>   b) white collar crime
>   c) pink collar crime
>   d) corporate crime

9. Which company used a formula in court to determine the cost of life versus the cost of fixing the problem?
>   a) General Electric
>   b) American Airlines
>   c) Ford Motor Company
>   d) IBM

10. Which of the following type of crime often develops among immigrants?
>   a) organized crime
>   b) white collar crime
>   c) pink collar crime
>   d) corporate crime

Ethical Debate Exercise

Divide the class into two groups and have each prepare for 15 minutes, allowing them to prepare for a class debate.

The debate over the death penalty has been complicated in recent years by questions regarding both the fairness of the criminal justice system and the possibility of reform and rehabilitation among death row inmates.

Group One:
The death penalty is NOT a deterrent and the costs associated with obtaining a death penalty conviction are larger than the costs associated with providing lifetime imprisonment.

Group Two:
The death penalty IS a deterrent and the costs associated with obtaining a death penalty conviction are less than those associated with providing lifetime imprisonment.

In Class Activity

As a group, fill in the worksheet below.

Beside each letter of the alphabet, come up with one social problem that begins with that letter.
A social problem = something you are concerned about & want changed.

A

B

C

D

E

F

G

H

I

J

K

L

M

N

O

P

Q

R

S

T

U

V

W

X

Y

Z

Outside Activity

Research the topic "Child Molestation" on the Internet using the following guidelines.

In order to complete this activity, you must log onto a computer. Please answer the following questions as thoroughly as possible.

1) Do a COMPLETE Sex Offender Search in your area. http://www.mapsexoffenders.com/ What or whom did you find, and where? Anything surprising? Attach a complete printout to this worksheet.

2) Log onto: http://www.pameganslaw.state.pa.us/

    a. Who was Megan?
    b. What does Megan's law intend to do?

Group Project Topics

Have each group member select one aspect of this research project.

Curtailing Drug Abuse: Internal Control or External Control?

Internal controls are those which exist within individuals.
External controls come from outside of individuals.

Attend a meeting of Alcoholics Anonymous or another support group that counsels alcohol or drug addiction.

Record observations watching for the following:
Organization of the meeting
How newcomers are welcomed
How members are taught to deal with their addiction
Other methods used to deal with addiction

Interview someone you know who has used drugs in the past and try to discover the extent to which each type of social control has been effective in curtailing the use of drugs.

Present the findings to the class.

Ideas for Outside Reading

Abadinsky, Howard. 2000. "Organized Crime." New York. Wadsworth.

Dabney, Dean. 2003. "Crime Types." Belmont, Ca. Wadsworth.

Hollinger, Richard. 1983. "Theft by Employees." Lexington, Mass. Lexington Books.

Meier, Robert, F. 1984. "Major Forms of Crime." Beverly Hills, Ca. Sage.

Rothman, David. 1990. "Perfecting the Prison: United States 1789-1865," in *Oxford History*, pp. 100-116 (CP)

CHAPTER EIGHT – GLOBAL STRATIFICATION
Video Clip Exercise

Watch the recommended video segments and then answer the questions below.

Video Segment 1:
Wadsworth's Lecture Launchers for Introductory Sociology:
"Life Chances: The Guerry Family"

Video Clip Exercise:
1. What are life chances?

2. How do life chances change for the Guerry family?

3. What are the pros and cons of self-employment?

Case Study Exercise

Visit this Internet Case study and then answer the following questions.

Poverty Facts and Stats

http://www.globalissues.org/TradeRelated/Facts.asp

Case Study Exercise:

(1) How many people live on less than two dollars a day?

(2) The GDP (Gross Domestic Product) of the poorest 48 nations (i.e. a quarter of the world's countries) is less than the wealth of the world's _____ richest people combined.

(3) Nearly _____ people entered the 21st century unable to read a book or sign their names.

(4) The _____ nation on Earth has the widest gap between rich and poor of any industrialized nation.

(5) The lives of _____ children will be needlessly lost this year because world governments have failed to reduce poverty levels

(6) According to UNICEF, _____ children die each day due to poverty

Group Quiz Exercise

Team up in groups to see who can answer the most correctly!

1. Slavery:
     a. has been rare in human history
     b. has usually been based on racism
     c. has varied in its condition from place to place
     d. was most common in pastoral societies

2. All except one of the following statements are true concerning slavery in America. Which does NOT belong?
     a. slavery originally included Hispanics
     b. slavery led to racism
     c. slavery became inheritable
     d. slavery failed among Indians

3. In a(n) _____ society people's status is determined by birth:
     a. class system
     b. caste system
     c. endogamous system
     d. rigid system

4. In a caste system, status is:
     a. reserved for the Brahmans
     b. achieved
     c. actively sought
     d. ascribed

5. In the Indian caste system, the merchants and skilled artisan caste is known as:
     a. Kshatriya
     b. Vaishva
     c. Brahman
     d. Shudra

6. In the United States, slavery was replaced with:
     a. a system of indentured servitude
     b. a racial caste system
     c. greater racial exploitation
     d. neopaternalism

7. A basic difference between class and caste stratification is that:
     a. social mobility is possible in class systems
     b. there is no inherited wealth in caste systems
     c. poverty is usually greater in caste systems
     d. in caste systems it is possible to "marry up" more than one caste level

8. According to Karl Marx, farmers and peasants are not considered social classes because they lack:
        a. class consciousness
        b. goals
        c. wealth
        d. status

9. _____ contended that social class is comprised of property, prestige, and power.
        a. Marx
        b. Weber
        c. Mills
        d. Moore

10. Who set forth the functionalist argument for why stratification is inevitable?
        a. Davis and Moore
        b. Harrington
        c. Galbraith
        d. Tumin

Ethical Debate Exercise

Divide the class into two groups and have each prepare for 15 minutes, allowing them to prepare for a class debate.

There has always been a distinction in the public's mind between the deserving poor and the undeserving poor.

Group One:
The deserving poor lie outside of the United States and we should lend financial aid to those within third world countries before lending aid to those within the borders of wealthy nations.

Group Two:
The deserving poor lie within the confines of the United States and we should lend aid to the poor within our borders before we lend aid to third world countries.

In Class Activity

http://www.cis.hut.fi/research/som-research/worldmap.html

Visit this website and answer the questions about global poverty below.

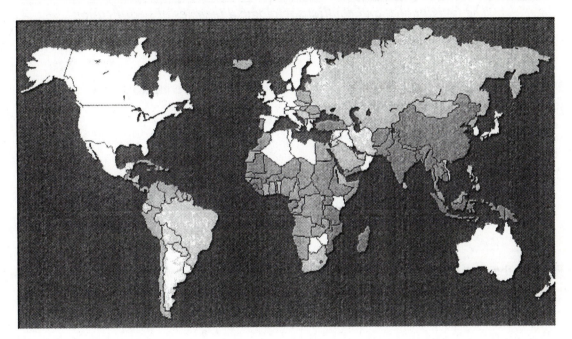

(1) The United States shares similar poverty levels with what other country?

(2) China shares similar poverty levels with what other country?

(3) Mexico shares similar poverty levels with what other countries?

Outside Activity

Research the topic "Sociological Theories of Stratification" using the following guidelines.

*The functionalist and conflict sociological theories of stratification
are very different.*

(1) Review the sections in the text highlighting functionalist and conflict theory. Summarize these findings in your own words.

(2) Review the sections in the text highlighting global stratification. Summarize these findings in your own words.

(3) Explain the structural functionalist theory of stratification in your own words.

(4) Explain the conflict theory of stratification in your own words.

Group Project Topics

Have each group member create a photo journal answering each of the following questions.

How Much Social Mobility Has Occurred In Your Family?

Vertical Social Mobility changes upward or downward in class or status between generations.
Intergenerational Mobility changes upward or downward from grandparents to parents.
Intragenerational Mobility changes upward or downward within your own lifetime.

**Provide family photo examples if available:
(1) Read the section on social mobility in the chapter on social stratification in your introductory text. In your journal, identify and explain:

> The structural factors that can affect social mobility
> Individual factors that can affect social mobility
> The way in which opportunities and life chances affect mobility

(2) Use the information above to trace, discuss, and analyze the intergenerational mobility within your family for at least two generations.

(3) Use the information above to trace, discuss, and analyze the intragenerational mobility within your family.

(4) Conclude your journal with a discussion of the relevance of historical events and mobility.

Ideas for Outside Reading

Basirico, Laurence, A. 1993. "Sociology in Action: Activities for Students." *Harper Collins College Publishers*. New York, NY: Harper Collins.

Brym, Robert J. and Lie, John. 2007. "Sociology." Thomson Wadsworth. Belmont, CA: Wadsworth.

Engels, Frederick. 1986. *The Origin of the Family, Private Property and the State*. New York: Penguin Books.

Fowler, Lori, A. 2007. "Introduction to Sociology: A Real World Approach." *Wadsworth*. Belmont, Ca: Wadsworth, Inc.

Weber, Max. 2002. *The Protestant Ethic and the Spirit of Capitalism and Other Writings*. New York: Penguin Books.

CHAPTER NINE – SOCIAL CLASS IN THE UNITED STATES
Video Clip Exercise

Watch the recommended video segments and then answer the questions below.

Video Segment 1:
Wadsworth's Lecture Launchers for Introductory Sociology:
"Effects of Social Mobility: A Personal Journey"

Video Clip Exercise:
1. In Elaine Bell Kaplan's life she has risen to what level through upward mobility?

2. What was the major factor that made upward mobility possible for Elaine Bell Kaplan?

3. What are some indicators of Elaine Bell Kaplan's socio economic status as a child and today?

Video Segment 2:
Wadsworth's Sociology: Core Concepts:
"Social Stratification"

Video Clip Exercise:
1. What are the major social classes in the United States?

2. In which class is inherited wealth most important?

3. How does one's education and occupation relate to one's social class?

4. How does television shape our ideas of social class?

Case Study Exercise

Use this table to answer the questions below.

Poverty Facts and Stats

Median Income of Households by Selected Characteristics and Income Definition: 2003 and 2004

(Households as of March of the following year)

Characteristic	2003 Median (2004 dollars)	2004 Median	Percent change
All households.	**44,483**	**44,389**	**-0.2**
Type of Household			
Family households...........................	55,442	55,327	-0.2
Married-couple families....................	64,082	63,813	-0.4
Male householder, no wife present.............	43,086	44,923	*4.3
Female householder, no husband present....	30,095	29,826	-0.9
Nonfamily households....................	26,433	26,176	-1.0
Male householder.........................	32,786	31,967	*-2.5
Living alone..........................	28,176	27,357	*-2.9
Female householder..........................	21,886	21,797	-0.4
Living alone..........................	19,178	19,446	1.4

Case Study Exercise:

(1) What is the dollar difference in median income for the years 2003 and 2004?

(2) Which type of family household reports the greatest amount of median income in 2004?

(3) Which type of family household reports the least amount of median income in 2004?

(4) What is the dollar amount difference between a female householder living alone and a married couple family?

Group Quiz Exercise

Team up in groups to see who can answer the most correctly!

1. In the US today, the median income is _____?
   a) $26,459
   b) $35,939
   c) $41,245
   d) $15,976

2. The richest people own their wealth in_____?
   a) rental properties
   b) bonds
   c) stocks
   d) trusts

3. How many people are classified as poor in this country?
   a) 22.1 million
   b) 35.7 million
   c) 41.7 million
   d) 50.9 million

4. What is wrong with the Government's Poverty Line Index?
   a) it does not take into account regional differences
   b) it is outdated
   c) it does not include all current communities
   d) it does not increase with current income trends

5. Eight out of ten college students have their own credit cards. This is the reason so many college students are in debt. How many credit cards does the average college student carry?
   a) One
   b) Two
   c) Three
   d) Four

6. 80% of the homeless are single mothers.
   a) True
   b) False

7. Homeless people have the right to sleep in public.
   a) True
   b) False

8. Homeless people do not have the right to vote.
   a) True
   b) False

Ethical Debate Exercise

Divide the class into two groups and have each prepare for 15 minutes, allowing them to prepare for a class debate.

"American society is full of opportunity."

Group One:
Opportunities for mobility are not abundant, and working full-time on minimum wage will not allow one to move out of the working poor today. The government SHOULD play an active role in reducing inequality in the United States.

Group Two:
Opportunities for mobility are abundant and it is up to the individual to make something of those opportunities through effort. The government should NOT play an active role in reducing inequality in the United States.

In Class Activity

Surviving on a Single Family Household Budget

You are a single parent with two children. Your children are ages 2 and 4 years old. You work full-time, 9am-5pm, 365 days a year. You never have a single day off!

(1) How much money would you earn on minimum wage, in your state for the year?

TOTAL INCOME: $_____

(2) Create an expense budget. Remember to include:

Clothing
Rent/House Payment
Daycare
Transportation
Car Insurance
Gas
Groceries
Healthcare
Diapers
Entertainment
Utilities (Gas, Electric, Water, Trash, Cable)
Telephone

TOTAL EXPENSES: $_____

(3) In order to survive you need to do without some excessive expenditures.
   Figure out how you can subtract some of these expenses in order to create a positive
   cash flow once again!

Outside Activity

Minimum Wage in the United States

(1) The current minimum wage in your state is $_____ per hour.

(2) Do you believe this minimum wage is too high, too low, or just right?

(3) According to your research, which state has the highest paying minimum wage rate?

(4) According to your research, which state has the lowest paying minimum wage rate?

Insert the minimum wage rates in the states below:

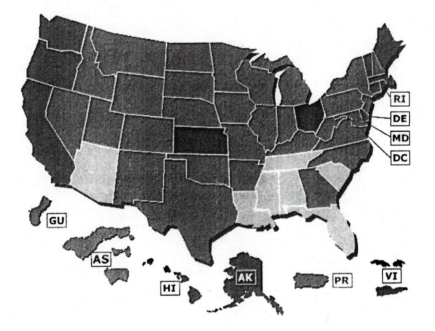

Group Project Topics

Have each group member analyze the affect of media and gender on attitudes toward welfare recipients.

Public Enemy Number One?

http://www.fair.org/index.php?page=1303

Visit this site and others like it to do research on the medias portrayal of who receives welfare benefits and government assistance.

(1) Is there a discrepancy in how men and women portray the poor in the media? Cite examples.

(2) Do the general public view men who receive aid differently from women who do? Who is seen in a more positive light?

(3) Do the media portray women with children differently than singles, or childless couples?

Combine all group findings and place them into a PowerPoint demonstration for the entire class.

Ideas for Outside Reading

Basirico, Laurence, A. 1993. "Sociology in Action: Activities for Students." *Harper Collins College Publishers.* New York, NY: Harper Collins.

Brym, Robert J. and Lie, John. 2007. "Sociology." Thomson Wadsworth. Belmont, CA: Wadsworth.

Engels, Frederick. 1986. *The Origin of the Family, Private Property and the State.* New York: Penguin Books.

Fowler, Lori, A. 2007. "Introduction to Sociology: A Real World Approach." *Wadsworth.* Belmont, Ca: Wadsworth, Inc.

Weber, Max. 2002. *The Protestant Ethic and the Spirit of Capitalism and Other Writings.* New York: Penguin Books.

CHAPTER TEN – RACE AND ETHNICITY
Video Clip Exercise

Watch the recommended video segments and then answer the questions below.

Video Segment 1:
Wadsworth's Lecture Launchers for Introductory Sociology:
"Genocide: Mike Jacobs' Story"

Video Clip Exercise:
1. For Mike Jacobs, what were the lasting effects of the Holocaust?

2. How has Mike Jacobs' story provided new insights into the Holocaust?

3. What did Mike Jacobs do to assure that others do not remain ignorant of the atrocities of the Holocaust?

Video Segment 2:
Wadsworth's Lecture Launchers for Introductory Sociology:
"Native American Assimilation"

Video Clip Exercise:
1. What are the major characteristics of a nation?

2. What is dual citizenship as it relates to the Hummingbirds?

3. What are the differences between Edward and Wilson Hummingbird?

Map the Stats Exercise

PERCENT WHO WOULD NOT WANT MEMBERS OF ANOTHER RACE AS NEIGHBORS

172 NATIONS OF THE WORLD

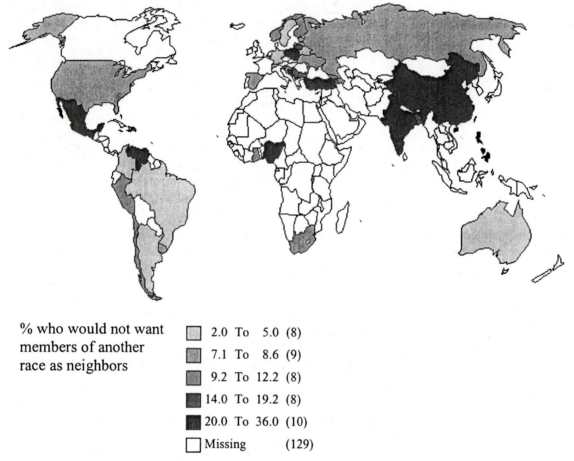

% who would not want
members of another
race as neighbors

☐	2.0 To 5.0 (8)
☐	7.1 To 8.6 (9)
☐	9.2 To 12.2 (8)
■	14.0 To 19.2 (8)
■	20.0 To 36.0 (10)
☐	Missing (129)

Map the Stats Exercise:

(1) Looking at this international map of the world, which countries appear to be most open to the idea of accepting other racial categories as neighbors? Which countries seem to be the least welcoming?

(2) Which do you think comes first: competition over resources, or racism?

Case Study Exercise

Holocaust Denial

---

http://en.wikipedia.org/wiki/Holocaust_denial

This exercise is about the history, development, and methods of Holocaust denial.

Case Study Exercise:

(1) _____ is the belief that the Holocaust did not occur as it is described by mainstream history.

(2) Over _____ were systematically killed by the Nazis and their allies.

(3) In addition, most Holocaust denial implies, or openly states, that the current mainstream understanding of the Holocaust is the result of a deliberate _____ to advance the interest of Jews at the expense of other peoples.

(4) Holocaust denial is also _____ in a number of _____ countries.

Group Quiz Exercise

Team up in groups to see who can answer the most correctly!

1. Which of the following are a social category comprised of those in society who are dominant in power, prestige, wealth, and culture?
    a) a minority group
    b) a majority group
    c) an affinity group
    d) the male species

2. The most powerful political positions under any political administration are of _____ descent.
    a) Irish
    b) German
    c) English
    d) French

3. Which of the following is a hostile attitude toward a person who belongs to a group, simply because he/she belongs to that group?
    a) stereotype
    b) prejudice
    c) discrimination
    d) racism

4. Which of the following describes what occurs when financial institutions mark off certain areas with a red pen and refuse loans to this area?
    a) de facto segregation
    b) reverse discrimination
    c) stratification
    d) red line districting

5. Which of the following represents when groups come together to take on characteristics they think are the best of both groups? The Melting Pot theory?
    a) blending theory
    b) multicultural education
    c) amalgamation
    d) accommodation

6. Which of the following occurs when a group tries to maintain uniqueness without conflict?
    a) blending theory
    b) multicultural education
    c) amalgamation
    d) accommodation

7. The earliest recorded history of multiracial people was in 1691.
_____ _____, an English white woman, was arrested and enslaved for the sole reason of having two children born of a black man.
     a) Susie Phipps
     b) Anne Wall
     c) Judy Harper
     d) Susan Brownmiller

8. In the Texaco discrimination case, executives were accused of what?
     a) Using racial slurs against employees, calling them "black jelly beans"
     b) Refusing to offer health insurance benefits to minorities
     c) Refusing to hire minorities
     d) None of the above

9. According to the Denny's lawsuit, three minority individuals experienced humiliation while trying to purchase a _____?
     a) Side of onion rings
     b) Hamburger
     c) Milkshake
     d) Sandwich

10. "Revisionists" try to rewrite academic history, and claim that the holocaust never happened.
     a) True
     b) False

Ethical Debate Exercise

Divide the class into two groups and have each prepare for 15 minutes, allowing them to prepare for a class debate.

In Focus: The Immigration Debate

Group One:
Immigration is a net benefit to the U.S. economy. Immigrants fill jobs that U.S. citizens often reject, help the U.S. economy maintain competitiveness in the global economy, and stimulate job creation in depressed neighborhoods. Immigrants contribute $25-30 billion more in taxes than they receive in services.

Group Two:
Employers hire foreigners who often work harder for less pay than U.S. citizens. Illegal immigration drained $51 billion more in social welfare and job displacement costs than immigrants paid in taxes.

In Class Activity

Experience Empathy Designing an In-Class Interview

Design a paper and pencil questionnaire with a fellow student who shares your racial or ethnic background, and then interview a student of another racial background in an effort to gain a sense of awareness.

Outside Activity

Is Race Still an Issue of Concern in Our Society Today?

Answer the following questions using the World Wide Web.

1) What types of hate groups exist? Why is the internet such an important tool for hate?

2) Visit the following website:
http://www.naacp.org/
What did you learn by visiting this site?

3) Visit this website:
http://sun3.lib.uci.edu/~dtsang/aware.htm
What did you discover here?

4) Visit this website:
http://www.tpronline.org/articles.cfm?articleID=153
What is important about this website?

5) Visit this website:
http://www.nativeamericans.com/
What is the legend of white buffalo?

6) Visit a website related to your ethnic or racial heritage: Why would this be a good resource for others?

7) Research exogamy and endogamy on the web. Which groups are most likely to practice each?

Group Project Topics

Each group member will analyze the Fair Housing Laws in your area.

Does Fair Housing Exist in Your Neighborhood?

In the United States, housing discrimination on the basis of age, gender, race, ethnicity, religion, physical disability, or type of family is against the law.

Fair Housing has been mandated by Title VIII of the Civil Rights Act of 1968

Have each group member select one of the following types of housing discrimination and research whether they are still prevalent, even though prohibited.

(1) **Steering**: when a real estate broker directs or steers minority group clients away from white areas, and steers white clients away from minority areas.

(2) **Intimidation**: a real estate broker warns clients by telling them that they probably don't want to live in a certain neighborhood because the residents are prejudiced against minorities.

(3) **Misrepresentation**: when minority clients are told that the property they are interested in has been sold when it has not.

(4) **Blockbusting**: when brokers try to frighten homeowners into selling their homes at very low prices by telling them that minorities are moving into their neighborhoods.

(5) **Redlining**: when mortgage lenders make it impossible to obtain a mortgage in ethnically mixed areas.

Ideas for Outside Reading

Basirico, Laurence, A. 1993. "Sociology in Action: Activities for Students." *Harper Collins College Publishers.* New York, NY: Harper Collins.

Brym, Robert J. and Lie, John. 2007. "Sociology." Thomson Wadsworth. Belmont, CA: Wadsworth.

Fowler, Lori, A. 2007. "Introduction to Sociology: A Real World Approach." *Wadsworth.* Belmont, Ca: Wadsworth, Inc.

Kornblum, William. 1997. "Sociology in a Changing World." Fort Worth, TX. Harcourt Brace College Publishers.

CHAPTER ELEVEN – SEX AND GENDER
Video Clip Exercise

Watch the recommended video segments and then answer the questions below.

Video Segment 1:
Wadsworth's Lecture Launchers for Introductory Sociology:
"Gender Inequality"

Video Clip Exercise:
1. What gender inequalities still exist today?

2. What is the glass ceiling? How can individuals break through the glass ceiling?

3. Are there other forms of glass ceilings that are faced by those other than women?

Map the Stats Exercise

% OF ADULT LABOR FORCE WHO ARE WOMEN

172 NATIONS OF THE WORLD

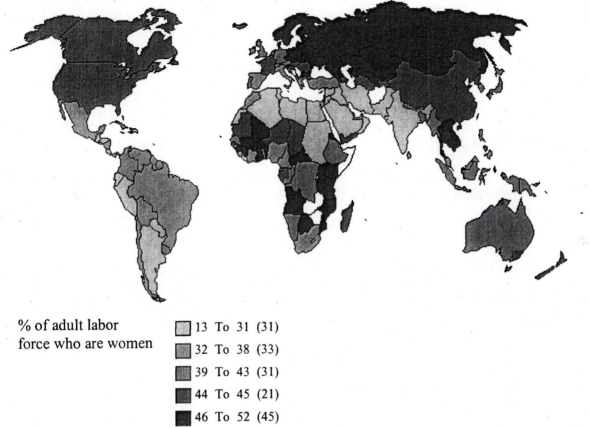

% of adult labor
force who are women

13 To 31	(31)
32 To 38	(33)
39 To 43	(31)
44 To 45	(21)
46 To 52	(45)
Missing	(11)

Map the Stats Exercise:

(1) Look at this map of the world and identify those countries with the highest participation among women in the labor force.

(2) Which countries have the lowest participation among women? Where does the United States fit in?

Case Study Exercise

The Glass Ceiling

http://www.inmotionmagazine.com/glass.html

This exercise is about the history, development, and perpetuation of the glass ceiling.

Case Study Exercise:

(1) _____ was Special Assistant to the Secretary of Labor, Robert Reich (during the Clinton administration). She was executive director for the Glass Ceiling Commission and Greenberg-Lake.

(2) We _____ live in a color blind or gender blind society. Sexism, racism, and xenophobia live side-by-side with unemployment, underemployment and poverty; they feed on one another and perpetuate a cycle of unfulfilled aspirations among women and people of color.

(3) Of the _____ percent of these managers who are women, only _____ percent are minority women.

Race	% of total workforce	Male	Female
non-Hispanic white	78.8%	43.2%	35.6%
African American	10.1%	4.1%	5.3%
American Indian / Eskimo / Aleut	0.6%	0.3%	0.3%
Asian & Pacific Islander	2.8%	1.4%	1.3%
Hispanic (of "white" and "other" races)	7.8%	4.6%	3.2%
non-Hispanic other	0.0%	0.0%	0.0%
	100.0%	54.3%	45.7%

(4) Who make up the greatest percentage of the total workforce? Who make up the least percentage of the total workforce?

Group Quiz Exercise

Team up in groups to see who can answer the most correctly!

1. Which of the following describe a person's biological maleness or femaleness?
     a) sex
     b) gender
     c) birth order
     d) menanglia theory

2. Who studied breast development?
     a) Karl Marx
     b) Emile Durkheim
     c) Susan Brownmiller
     d) Eve Steormann

3. Which of the following represent how society expects you to perform your gender the way you "ought to?"
     a) Socialization
     b) Gender rearing
     c) Gender accomplishment
     d) Color coding

4. Women are less satisfied with their bodies than are men.
     a) True
     b) False

5. Mattel is the world's 4th largest apparel manufacturer in the world.
     a) True
     b) False

6. The sociological significance of gender is that:
     a) it is a biological given that socialization cannot change
     b) it is one way society controls its members
     c) it provides new members for society
     d) it provides the most important social roles

7. Unequal access to power, property, and prestige on the basis of sex is found in:
     a) agricultural and industrial society
     b) every society except hunting and gathering
     c) only societies with agricultural surplus
     d) every society

8. After 1920, the first wave of the women's movement:
     a) dissolved
     b) turned away from its liberal wing
     c) was influenced by communists
     d) fought for Social Security

Ethical Debate Exercise

Divide the class into two groups and have each prepare for 15 minutes, allowing them to prepare for a class debate.

Illinois Rape Law: Illinois Gov. Rod Blagojevich tightened rape legislation in July, 2003. Under the law, if an individual says "no" at any time during the sexual act, the other person must stop or it becomes rape.

The case centered on a 17-year-old boy and whether he could be charged with rape after a girl consented to sex during a party, but changed her mind during the act and he refused to stop immediately.

A court decided that once consent was given, someone could not be punished for continuing sex. The CA Supreme Court disagreed and allowed the boy to be charged and later convicted.

Group One:
The law is important to make it clear to victims, offenders, prosecutors and juries that people have the right to halt sexual activity at any time. This Illinois Rape Law sends messages to potential rapists and will help encourage more individuals to come forward when they believe they have been raped.

Group Two:
The Illinois Rape Law gives people the right to change their minds during consensual sex and this will lead to false sexual-assault accusations and difficult court cases.

In Class Activity

Body Image in the Media

1. Select a clothing ad out of a fashion magazine. Write down all of the immediate adjectives that come to mind when viewing this photo. Does your photo contribute to positv4e or negative body image? How so?

2. What do you believe to be the main message in the ad? What specific product is being marketed? What nuances are being advertised?

3. Do you believe sex sells? Why or why not? Is your ad using sex to sell? Is it working?

4. What would an 8 year old girl think or feel about this ad? Be specific.

5. Re-draw your ad in a healthy fashion below:

Outside Activity

Is Household Equality Between Men and Women Increasing?

*The family in which the male is the primary breadwinner
and the female is the homemaker is declining.*

Answer the following questions using field observation techniques.

1) Observe married couples with children at a pediatrician's office, a PTA meeting, a school bus stop, an elementary school, a grocery store, a playground, or any other area where parents and children find themselves. Observe at least two of these situations for a total of one hour. Record the observations in a journal.

2) Compare the way in which the responsibilities are divided among husband and wife.

3) Are the roles and responsibilities stereotypically divided or not? Cite examples in your field study.

4) Based on your own observations, do you feel equality has increased, decreased, or remained the same?

Group Project Topics

*Each group member will analyze the Feminist Movement from its
inception to its current state.*

The Feminist Movement

http://www.beingjane.com/?OVRAW=the%20feminist%20movement&OVKEY=feminist%20moveme
nt&OVMTC=standard

http://www.blythe.org/peru-pcp/docs_en/feminist.htm

Feminism: the belief in the need to secure, or a commitment to securing, rights and opportunities for women equal to those of men.

Be sure to include the following in your research project:
> The History of Feminism
> The Reaction to Feminism
> The Waves of Feminism
> The Men's Movement

Ideas for Outside Reading

Basirico, Laurence, A. 1993. "Sociology in Action: Activities for Students." *Harper Collins College Publishers.* New York, NY: Harper Collins.

Brym, Robert J. and Lie, John. 2007. "Sociology." Thomson Wadsworth. Belmont, CA: Wadsworth.

Fowler, Lori, A. 2007. "Introduction to Sociology: A Real World Approach." *Wadsworth.* Belmont, Ca: Wadsworth, Inc.

Osborne, Richard. 2002. "Introducing Sociology." *Icon Books, Ltd.* UK, Europe: Macmillan Distribution Ltd.

Ruane, Janet, M. 1997. "Seeing Conventional Wisdom through the Sociological Eye." *Pine Forge Press.* Thousand Oaks, Ca: Sage.

CHAPTER TWELVE – AGING AND INEQUALITY BASED ON AGE
Video Clip Exercise

Watch the recommended video segments and then answer the questions below.

Video Segment 1:
Wadsworth's Lecture Launchers for Introductory Sociology:
"Then and Now: Aging in the Movies"

Video Clip Exercise:
1. What are the images that the media presents of the elderly? Are they positive or negative?

2. What are some stereotypical images that we have regarding the elderly?

3. Have images of the elderly as seen in movies changed over the past 60 years?

Map the Stats Exercise

% LIVING IN NURSING HOMES

50 STATES

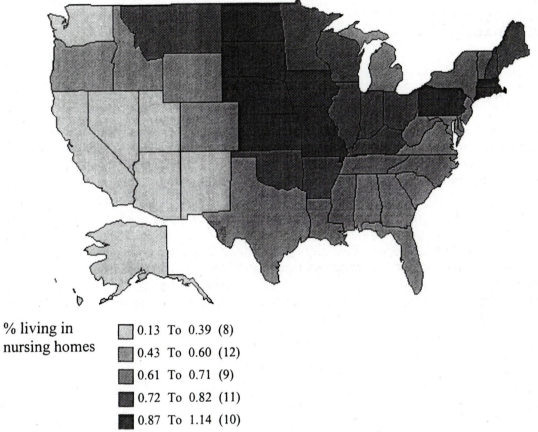

% living in nursing homes		
☐	0.13 To 0.39	(8)
▨	0.43 To 0.60	(12)
▩	0.61 To 0.71	(9)
▦	0.72 To 0.82	(11)
■	0.87 To 1.14	(10)

Map the Stats Exercise:

(1) According to the map, the majority of nursing home citizens live in which part of the country?

(2) Where do the least number of nursing home residents live? Why do you think this is so?

Case Study Exercise

Case Studies: Is This Elder Abuse?

Case Study Exercise:

Case 1

Is This Elder Abuse?

Sue, 70, a widow, lived in a small apartment with her son, Jason, 44. Jason had been in and out of drug and alcohol treatment centers for years, but was doing well for the last six months since he moved back in with his mother. Sue knew here son had nowhere else to go so she took him in under two conditions: he had to find a job and he could not drink.

Jason found a job and things seemed to be going well until he stopped coming home right after work. Sue knew he was stopping at the corner bar because she could smell the alcohol on his breath. The third time this happened, Sue confronted her son. Jason immediately became belligerent, verbally abusing her and forcing her to go to her room. The next night Sue confronted him again threatening to throw him out if he continued to drink. Jason became enraged and started running toward his mother with his fist raised over his head. Fearing for her life, Sue fled to the safety of her neighbor's house.

Case 2

Is This Elder Abuse?

Carol, 24, divorced, lived on the second floor of an apartment with her two young children. Living below her on the first floor was Beatrice, 86, a nice old lady who didn't leave her apartment very often because of her arthritic knees and poor eyesight. Carol and her children visited Beatrice frequently and often helped with her laundry in exchange for occasional babysitting. Beatrice loved their company.

Every Saturday, Carol offered to do the grocery shopping for Beatrice. Because she could not do it herself, Beatrice was happy to accept Carol's help. Carol thought it was okay to keep $20 of the change each week because she was taking the time and trouble to help Beatrice; although, she was never offered any money. Carol thought Beatrice would never realize the money was missing because of her poor eyesight.

Group Quiz Exercise

Team up in groups to see who can answer the most correctly!

1. The young old are composed of the frail and sick aged.
    a) True
    b) False

2. Geriatrics studies the process of aging.
    a) True
    b) False

3. The Programmed Aging theory plays more of a role than the Wear and Tear Theory of Aging.
    a) True
    b) False

4. All except one of the following are reasons people live longer in Industrial societies. Which one should not belong?
    a) fighting disease effectively
    b) having a plentiful food supply
    c) purified water
    d) stressing physical fitness

5. The life span in highly industrialized nations has:
    a) decreased, then increased
    b) increased slightly
    c) not changed at all
    d) increased sharply

6. Prejudice, discrimination, and hostility toward older people are known as:
    a) ageism
    b) anti-aging
    c) age stereotyping
    d) age stigma

7. One sign that a shift in the meaning of old age has occurred is that:
    a) people lie about their age
    b) special ministers re appointed to assist the elderly
    c) children spend more time helping the elderly
    d) children now sign support agreements

8. Social Security arose out of the conflict over:
    a) retirement
    b) the New Deal
    c) the Alphabet War
    d) the Townsend Plan

Ethical Debate Exercise

Divide the class into two groups and have each prepare for 15 minutes, allowing them to prepare for a class debate.

One of the most important public policy debates today surrounds the issues of euthanasia and assisted suicide. The outcome of that debate will profoundly affect family relationships, interaction between doctors and patients, and concepts of basic ethical behavior.

Group One:
There are terminally ill patients who wish to die with dignity. They are in such pain that euthanasia may be the most humane thing to do.

Group Two:
There are many examples of people wanting to die, but after getting through the pain, they realize what a mistake euthanasia might have been. New medicines relive pain better than ever before.

In Class Activity

The Aging Population is Booming!

The increasing number of elderly will affect how people perform their jobs and the opportunities for employment in various professions. Identifying these occupations and examining them will be useful.

1. If you are currently an unemployed, full-time student, how do you see the college campus being affected by increasing numbers of elderly individuals? If you are currently working, how will your specific industry be impacted by increasing numbers of elderly within the population?

2. How will medicine and the health care industry be impacted by the increase in the aged population?

3. How will law enforcement be impacted by the increase in the aged population?

4. A large developer is planning a master community in your area. How can your sociological knowledge be beneficial to their planning and design of this master community? What should you tell them to account for concerning the aged population?

5. Draw the perfect community plan below:

Outside Activity

Using Sociological Theories to Examine Social Issues Facing the Elderly

Gerontology has become one of the most rapidly growing areas of sociology. There are many social problems the elderly face a result of the rapidly increasing elderly population in the United States.

Answer the following questions using Sociological Theory.

1) Use the **structural functional** theories to research depression and suicide among widowers. How might functionalism help explain this phenomenon?

2) Use **symbolic interaction** theories to research stereotypes regarding the aged, ageism, and how the aged feel about them. How might symbolic interactionism help explain ageism?

3) Use **conflict theory** to examine forced retirement, discrimination in the workplace, and diminishing social security benefits. How might conflict theory help explain these trends?

Group Project Topics

Each group member will analyze the treatment of the elderly from a cross-cultural perspective.

The Treatment of Elderly around the Globe
Each group needs to creatively explore the treatment of elderly populations in other cultures.

You can conduct interviews, create a video, design a paper and pencil questionnaire, or design a PowerPoint presentation.

Be sure to include the following in your research project:
1- Roles of the elderly within the family
2- Roles of the elderly within the larger community
3- How are the elderly portrayed in the media?
4- To what extent is ageism a problem?
5- Levels of respect
6- Is aging a valued process or a feared process?
7- Are the elderly treated with equality?
8- Are there any cultural traditions prevalent surrounding the elderly population?

The ultimate goal is to compare the treatment of the elderly population in the United States to that of elderly populations in other cultures.

Ideas for Outside Reading

Basirico, Laurence, A. 1993. "Sociology in Action: Activities for Students." *Harper Collins College Publishers*. New York, NY: Harper Collins.

Brym, Robert J. and Lie, John. 2007. "Sociology." Thomson Wadsworth. Belmont, CA: Wadsworth.

Fowler, Lori, A. 2007. "Introduction to Sociology: A Real World Approach." *Wadsworth*. Belmont, Ca: Wadsworth, Inc.

Osborne, Richard. 2002. "Introducing Sociology." *Icon Books, Ltd.* UK, Europe: Macmillan Distribution Ltd.

Ruane, Janet, M. 1997. "Seeing Conventional Wisdom through the Sociological Eye." *Pine Forge Press*. Thousand Oaks, Ca: Sage.

CHAPTER THIRTEEN – THE ECONOMY AND WORK IN GLOBAL PERSPECTIVE
Video Clip Exercise

Watch the recommended video segments and then answer the questions below.

Video Segment 1:
Wadsworth's Lecture Launchers for Introductory Sociology:
"Globalization"

Video Clip Exercise:
1. Our world has become more globalized and this has resulted in environmental challenges. Give some examples.

2. How does technology relate to globalization?

3. How does the struggle between globalization and technology challenge some of our values?

Video Segment 1:
Wadsworth's Sociology: Core Concepts
"The Economy"

Video Clip Exercise:
1. What are the basic needs that are met by the institution of economics?

2. What markets are changed as a result of globalization?

3. How have technology and robotics changed the production of goods and services?

Case Study Exercise

China: The Last Frontier
http://www.essentialaction.org/addicted/country.html

For the cigarette companies, the Chinese market represents the proverbial mother lode, a potential savior from declining sales at home. One out of every three cigarettes smoked in the world today is smoked in China.

Case Study Exercise:

1) What projection regarding smoking deaths in China was recently made by the American Medical Association?

2) Why is China the target for large tobacco companies?

3) Describe what has happened with the large cigarette companies recently.

4) Describe how smuggling beneficial to large cigarette manufacturers.

Group Quiz Exercise

Team up in groups to see who can answer the most correctly!

1. Most people in postindustrial society would work in the _____ sector of the economy.
   - a. primary
   - b. service
   - c. tertiary
   - d. quadiary

2. All except one of the following represent wealth in the United States EXCEPT one. Which does not belong?
   - a. the gap between the rich and poor is higher than it has been in generations
   - b. the richest 5th receive half of all income in the United States
   - c. the middle class has fallen to the bottom 50% of the class scale
   - d. the poorest 5th receive less than 5% of all income

3. In laissez-faire capitalism, government does not interfere with:
   - a. legal company agreements
   - b. market forces
   - c. international competition
   - d. tariff restrictions

4. The current form of the United States economy is:
   - a. democratic capitalism
   - b. pure capitalism
   - c. postindustrial capitalism
   - d. welfare capitalism

5. A key feature of socialism is:
   - a. repression
   - b. market forces
   - c. party membership
   - d. central planning

6. Capitalism and socialism each see each other as:
   - a. a system of exploitation
   - b. a threat to state security
   - c. a threat to social order
   - d. an antidemocratic system

7. Convergence theory states:
   - a. future economies will be a mixture of socialism and capitalism
   - b. multinational corporations are blurring national borders
   - c. the world is moving towards a single international government
   - d. the gap between rich and poor is growing smaller

8. To indicate how modern companies dominate the economy, sociologists use the term:
   a. international cartels
   b. multinational corporations
   c. interlocking directorates
   d. corporate capitalism

Ethical Debate Exercise

Divide the class into two groups and have each prepare for 15 minutes, allowing them to prepare for a class debate.

**Is Global Income Inequality Increasing or Decreasing?
Does global integration make the poor richer or poorer?**

Group One:
Milanovic (1999) put together data on a global scale using household surveys and found an important increase in inequality caused by slower growth of rural incomes in Asian countries compared to the rich countries.

Group Two:
Sala-i-Martin has generated considerable media interest with a paper which suggests that income inequality is actually falling. He has generated a different data set based on aggregate income and estimates of within country distributions of income between rich and poor. This data set suggests that global income inequalities were falling.

In Class Activity

What is That Job Worth?

Occupation	Median Income
Accountant	$47,000
Actor	$23,470
Actuary	$69,970
Advertising Manager	$57,130
Aerospace Engineer	$72,750
Anesthesiologist	$306,964
Architect	$56,620
Chemist	$52,890
Chiropractor	$65,330
Claims Adjuster	$43,020
College Instructor	$49,040
Computer and Information Systems Manager	$85,240
Computer Software Engineer	$70,900
Construction Manager	$63,500
Court Reporter	$41,550
Dentist	$123,210
Designer	$52,260
Elementary School Administrator	$71,490
Engineer	$90,930
Farmer	$43,740
Funeral Director	$43,380
General Surgeon	$255,438
Gynecologist	$233,061
Human Resources Manager	$64,710
Judge	$94,070
Lawyer	$90,290
Medical Scientist	$56,980
Musician	$36,290
Photographer	$24,040
Physical Therapist	$57,330
Pre-School Teacher	$39,810
Property Manager	$36,880
Property Surveyor	$42,870
Psychologist	$51,170
Psychiatrist	$163,144
Registered Nurse	$48,090
Social Scientist	$52,280
Social Worker	$33,150
Top Executive	$68,210

All Data from: http://www.bls.gov/oco/ocos074.htm

> **Have the students randomly select one of the above occupations and then answer the following questions. Assign each a racial category and family lifestyle as well.**

1) What is your occupation? What is your race? What is your family lifestyle?

2) Go to the following website: http://www.bls.gov/oco/ocos074.htm
    a) What is the nature of your work?

    b) What are your working conditions?

    c) What training is required for your occupation?

    d) Describe your job outlook?

3) What is your annual income, as listed in the chart above? Remember- gross Income is not net Income! What do you think your take home pay would be?

4) Based on this income, and your lifestyle, design a realistic expense budget:
    a) What type of house will you live in? What do you think this will cost?

    b) What type of car will you drive? What do you think this will cost?

    c) Do you have children? What ages? Are they in public or private school? Cost?

    d) What other expenses might you afford with your occupation?

    e) What can you not afford, that you would like to, because of your occupation?

    f) Based on your occupation, do you believe you have school loans that need to be paid off?

    g) If you could choose any occupation above, what would it be? Why?

Outside Activity

Research the World of Opportunity available to those with Sociology Degrees!

Careers in Sociology

Sociology can provide a rich fund of knowledge and many distinctive ways of looking at the world so as to generate new ideas and assess the old. Sociology can give undergraduates a basic standard preparation for career options in all the fields listed below, with advanced study and/or training in relevant work experience to qualify for full professional status.

Areas of Employment

Advertising:
Mass Communication Analysis, Motivational Research, Public Opinion Polls.

Banking and Finance:
Community Relations, Customer Relations, Employee Relations, Organizational Planning, Personnel Administration.

Communications:
Consumer Research, Labor Relations, Employee Relations, Mass Communications Analysis, Public Opinion Polls.

Criminal Justice:
Law Enforcement, Criminology, Delinquency Prevention, Penology and Corrections, Probation & Parole, Court Administration, Private Security.

Insurance:
Area Studies, Community Relations, Customer Relations, Employee Relations, Population Studies, Public Health.

Manufacturing:
Community Relations, Employment Interviewing, Customer Relations, Employee Relations, Industrial Sociology, Labor Relations, Motivational Research, New Business Research, Organizational Planning, Personnel Administration, Manpower Resource Studies.

Marketing:
Consumer Research, Motivational Research, Population Administration.

Merchandising:
Consumer Research, Customer Relations, Personnel Administration.

Medical Services:
Alcoholism & Drug Abuse, Addiction Research, Community Relations, Employee Relations, Hospital Casework, Medical Sociology, Organizational Planning, Public Health.

Social Services & Government:
Child Welfare, Social Work, Community Agency, Community Planning & Development, Community Organization, Family Welfare, Group Interaction, Housing Studies, Interviewing (non-diagnostic), Manpower Resource Studies, Minority Group & Race Relations Studies, Population Studies, Public Welfare-Housing, Casework, Race Relations, Urban Planning.

A Sociology major is also an excellent preparation for the study of law, public administration, hospital administration, publishing, rehabilitation counseling, recreation planning and administration, career planning and counseling, and sales.

Employment Outlook in Some Occupations Related to the Sociology Major

Urban Planning:
The rapid change in our way of life, urban development and sprawl, housing, traffic, land use, urban renewal and urban development all call for a broad understanding of the social forces at work. Previously, such planning placed heavy emphasis on engineering and scientific backgrounds, but not so today. In fact, urban planning has developed a professional program of studies leading to an advanced degree.

Social Work:
Social work, a professional method of helping people with diverse problems, may be entered directly from college, although graduate work is rapidly becoming essential for positions of leadership and responsibility. Social workers practice casework, group work, social research, social planning, health administration, and community organization.

Market and Survey Research:
Market and survey researchers are employed in advertising agencies, manufacturing companies, communications industries, and independent firms that conduct public opinion research. Individuals engaged in market and survey research departments determine how people vote and why; who watches what in the media and why; and who buys what product and why. Those who work in independent firms are often called as consultants to business to determine consumer values, attitudes of employees and management toward communication channels, attitudes toward government programs, and forecasting consumer and business trends.

Human Resources/Administration:
The increase in job discrimination suits, pension laws, federal regulations and labor disputes have made the human resources department an integral part of organizations (governmental, industrial, hospital, etc.). The tasks of people engaged in human resources/administration include recruiting and interviewing, employee counseling, testing of employee attitudes, wage and salary administration, labor relations, job development, and employee training. Human Resources/Administration have developed a professional program of studies leading to an advanced degree.

Medical Services/Administration:
Individuals engaged in Medical Services/Administration are employed by government departments (e.g., public health, mental health, food and drug), voluntary health agencies, planning agencies, mental health centers, hospitals, nursing homes, alcohol addiction & drug abuse centers, and senior citizen centers specializing in health, recreation, nutrition, education, housing and employment for the aging. Medical Services/Administration have developed a professional program of studies leading to an advanced degree (e.g. rehabilitation counseling and administration, gerontological services administration, health services administration, and community mental health).

What Can I Do With a Sociology Degree?
There are many careers out there for sociology majors. Below, you will find areas of job interest, employers who hire in that area, and strategies (including classes to take, experience to gain, interviewing, etc.) to get that job.

Areas	Employers	Strategies
Environment & Society	• Waste Management Firms • Health Agencies • Local Planning Agencies • Environmental Advocacy Groups • National Advocacy Groups • Environmental Periodicals • Federal Government • Regional, State, & Local Agencies • Consulting Firms	• Take courses in general and urban ecology, statistics, and public speaking. • Learn skills in communication networking, electronic mail, and analytical writing. • Gain experience via part-time or internships and volunteer work. • Obtain graduate degree for advancement.
Criminal Justice-- Corrections Rehabilitation Law Enforcement Judiciary	• Correction Institutions • Court Systems • Federal, State, & Local Government, • Law Enforcement Agencies	• Emphasize research methods, statistics and computer skills. • Gain essential practical experience via part-time, internships, & volunteer work. • Learn skills in communication networking & electronic mail. • Obtain graduate degree for advancement.
Demography	• Companies & Organizations doing demographic forecasting and population studies. • Companies compiling market research data. • Consulting firms • Business & Industry • International, Federal & State Agencies • Government & Regional Planning Departments • Colleges & Universities • Foundations • Advertising & Marketing Firms • Private Research Firms	• Take courses in social research methods, statistics, computer applications, population, social movements, calculus, & foreign language. • Develop good communication skills. • Gain essential practical experience via part-time, internships, & volunteer work. • Learn skills in communication networking & electronic mail. • Obtain graduate degree for advancement.

Human Services-- Counseling Advocacy Mental Health Services Social Services-- *--Administration* *--Programming* *--Recreation*	• Advocacy Groups • Federal, State, & Local Government • United Way Agencies/Local Branches of National Non-Profit Organizations • Religiously-Affiliated Service Organizations • Adoption & Child Care Agencies • Public & Private Nursing Homes • Hospitals	• Learn skills in communication networking & electronic mail. • Concentrate course work in area of interest. • Pursue excellent academic record. • Emphasize research methods, statistics and computer skills. • Obtain counseling courses and experience for counseling & caseworker positions. • Acquire related experience. • Obtain graduate degree for advancement. • Consider business minor or double major for positions in administration, & master's in health care administration for advancement. • May need master's degree in social work.
Business-- Demography/Planning Human Resources Management Sales Advertising Marketing Consumer Research Insurance Real Estate Personnel/Training Administration	• Research Departments/Firms • Personnel/Industrial Relations Departments • Marketing/Management Firms • Organizational Planning Departments/Firms • International Business • Manufacturing Firms • Advertising Firms • Consumer Research/Public Relations Firms • Insurance/Real Estate Companies • Publishing Firms • Consulting Firms	• Focus on an occupational area: personnel, industrial relations, management, marketing, or public relations. • Take courses in labor relations, industrial psychology, sociology of work, personnel management, public speaking, writing, social psychology, and human/social interaction. • Gain related experience. • Learn skills in communication networking & electronic mail. • Obtain graduate degree for advancement. • Stress work experience, social skills, public speaking, writing skills, statistics & research training when interviewing.
Education-- Teaching Administration Alumni Relations Placement Offices Research	• Public & Private Schools • Colleges & Universities	• Obtain certification/licensing to teach grades K-12. • Volunteer as a tutor. • Obtain Ph.D. to teach and for advanced research positions in colleges & universities. • Secure strong personal recommendations. • Learn skills in communication

		• networking & electronic mail. • Take courses in sociology of education, social psychology, developmental psychology,
Social Science Research-- --Market Research Analysis --Evaluation Research	• Federal Government • National Headquarters of Non-Profit Organizations • Firms conducting social, market, or statistical research • Public Relations Firms • Professional Periodicals • Newspapers & Magazines • Social Service Agencies • Hospitals • Business & Industry • Labor Unions • Professional Sociologists • Universities • Religious Organizations • Public Opinion Research Polls	• Learn federal government job application process. • Develop strong quantitative, statistical, writing, informational gathering, and assimilating skills. • Take courses in research methods and statistics. • Learn skills in communication networking & electronic mail. • Acquire related experience. • Obtain graduate degree for advancement and specialized areas.
Community Relations	• Federal, State, & Local Government • National headquarters & local branches of non-profit organizations • Private Social Service Organizations • Religiously Affiliated Service Organizations • Child Care Agencies	• Learn federal, state & local government job application process. • Obtain experience in counseling, advocacy or administration. • Acquire knowledge of community problems & government resources. • Gain volunteer experience. • Take courses in public speaking, inequality, social classes, race relations, gender, social psychology, social/human interaction. • Learn skills in communication networking & electronic mail.
Government-- Social Science Analysis Social Statistics Demography Administration Management Program Development Policy Analysis	• Public Assistance Agencies • Federal, State, Local, & County Government	• Learn federal, state & local government job application process. • Learn skills in communication networking & electronic mail. • Obtain experience in research & evaluation. • Acquire skills in survey & evaluation research, and specialties in such fields as

Personnel Research Investigation		medical/health sociology, aging, criminal justice, demography, and family. • Gain experience via co-op programs and volunteer work. • Obtain graduate degree for advancement.

"What can I do with a BA in sociology?" As a strong liberal arts major, sociology provides several answers to this important question:

- A BA in sociology is excellent preparation for future graduate work in sociology in order to become a professor, researcher, or applied sociologist.
- The undergraduate degree provides a strong liberal arts preparation for entry level positions throughout the business, social service, and government worlds. Employers look for people with the skills that an undergraduate education in sociology provides.
- Since its subject matter is intrinsically fascinating, sociology offers valuable preparation for careers in journalism, politics, public relations, business, or public administration--fields that involve investigative skills and working with diverse groups.
- Many students choose sociology because they see it as a broad liberal arts base for professions such as law, education, medicine, social work, and counseling. Sociology provides a rich fund of knowledge that directly pertains to each of these fields.

"What can I do with an MA or PhD degree in sociology?" With advanced degrees, the more likely it is that a job will have the title sociologist, but many opportunities exist--the diversity of sociological careers ranges much further than what you might find under "S" in the Sunday newspaper employment ads. Many jobs outside of academia do not necessarily carry the specific title of *sociologist*:

- Sociologists become high school teachers or faculty in colleges and universities, advising students, conducting research, and publishing their work. Over 3000 colleges offer sociology courses.
- Sociologists enter the corporate, non-profit, and government worlds as directors of research, policy analysts, consultants, human resource managers, and program managers.
- Practicing sociologists with advanced degrees may be called research analysts, survey researchers, gerontologists, statisticians, urban planners, community developers, criminologists, or demographers.
- Some MA and PhD sociologists obtain specialized training to become counselors, therapists, or program directors in social service agencies.

Group Project Topics

Each group member will analyze the treatment of the elderly from a cross-cultural perspective.

Multinational Corporations
Read the section about multinational corporations in your text.

Identify three multinational corporations that your group would like to learn more about. Select three different corporations which manufacture three different types of products.

Find 10 articles about these corporations that have appeared in financial magazines or newspapers over the last 24 months that will allow you to answer the following questions:
1- What types of national and international issues outside of the corporation affect how it might operate? (tariffs, foreign policy, laws).
2- Are there any specific company policies held within the corporation that you feel might affect family members?
3- Are there any specific company policies held within the corporation that you feel might affect communities?
4- Are there any specific company policies held within the corporation that you feel might affect specific states?
5- Are there any specific company policies held within the corporation that you feel might affect the nation?
6- Are there any specific company policies held within the corporation that you feel might affect other countries?
7- Do you feel the company is more influenced by outside policies, or do you feel the corporation is acting as a policy maker?

Ideas for Outside Reading

Basirico, Laurence, A. 1993. "Sociology in Action: Activities for Students." *Harper Collins College Publishers.* New York, NY: Harper Collins.

Brym, Robert J. and Lie, John. 2007. "Sociology." Thomson Wadsworth. Belmont, CA: Wadsworth.

Firebaugh, G. and B. Goesling (2004). "Accounting for the Recent Decline in Global Income Inequality." *American Journal of Sociology,* Volume 110 (Number 2 (September 2004)): 283–312.

Milanovic, B. (2002). True world income distribution, 1988 and 1993: First calculation based on household surveys alone, *Economic Journal* 112, 476.

Ravallion, M. 2003 "The debate on globalization, poverty and inequality: why measurement matters", Policy Research Working Paper 3038, World Bank.

CHAPTER FOURTEEN – POLITICS AND GOVERNMENT IN GLOBAL PERSPECTIVE
Video Clip Exercise

Watch the recommended video segments and then answer the questions below.

Video Segment 1:
Wadsworth's Lecture Launchers for Introductory Sociology:
"Integrity of the Electoral Process"

Video Clip Exercise:

1. Explain the importance of every eligible voter exercising his or her right to vote in light of the 2000 Presidential election.

2. How does the Electoral College work and is it truly representative of the population?

3. What are the implications for future elections?

Map the Stats Exercise

% OF WOMEN IN STATE LEGISLATURES

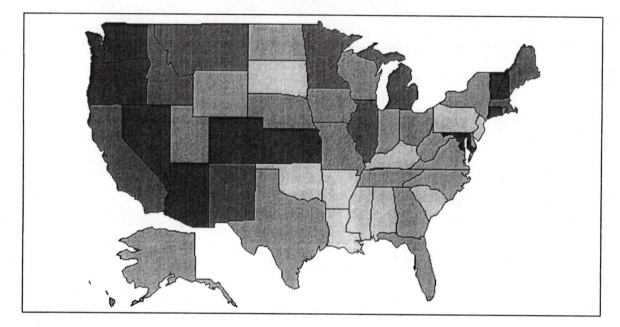

% of women in
state legislatures

- 8.0 To 16.2 (10)
- 16.5 To 19.5 (10)
- 20.6 To 24.5 (10)
- 24.7 To 28.6 (10)
- 29.3 To 40.8 (10)

Map the Stats Exercise:

(1) Which states have the highest percentage of women in state legislatures?

(2) Which states have the lowest percentage of women in state legislatures?

Case Study Exercise

GENDER GAP, GOP EDGE IN SMALL DONATIONS COULD LOOM BIG IN 2004 ELECTIONS

http://www.opensecrets.org/pressreleases/DonorDemog.asp

Women with incomes separate from their spouses gave most of their money to Democrats during the 2002 election cycle, while women who listed a non-income-earning occupation sent most of their contributions to Republicans.

Case Study Exercise:

1) This study was based on analysis of more than _____ contributions of _____ or more given during the 2001-2002 election cycle.

2) The study also found that _____ were far more reliant than _____ on deep-pocketed givers in the election cycle.

3) Looking at giving by gender, the Center found that women who _____ or _____ in their contributions during the 2002 election cycle gave 61 percent of their money to Democrats and 39 percent to Republicans.

4) Women who identified themselves as _____ or _____ occupation preferred Republicans over Democrats by 55 percent to 45 percent.

5) Among very wealthy donors, Democrats reigned supreme. Contributors of _____ or more gave 92 percent of their money to Democrats.

6) The findings illustrate the _____ strong advantage over _____ in the current system, which caps total contributions to candidates.

Group Quiz Exercise

Team up in groups to see who can answer the most correctly!

1. Legitimate power is known as:
    a. agency
    b. right of office
    c. authority
    d. authorization

2. All EXCEPT one of the following are types of authority proposed by Weber. Which one should NOT be included?
    a. organizational
    b. rational-legal
    c. traditional
    d. charismatic

3. Charismatic leaders often face opposition in traditional and ration-legal societies because:
    a. their groups are inherently unstable
    b. there are no firm rules of succession
    c. they usually oppose the established order
    d. they pose a threat to the established order

4. The development of surpluses around 3500 BC caused:
    a. large horticultural societies to emerge
    b. absolute rulers to dominate city-states
    c. the lessening of social inequalities
    d. cities to evolve

5. The idea of everyone having the same basic rights because they are born in a country is termed _____ citizenship
    a. common
    b. universal
    c. natural
    d. inherent

6. The age group least likely to vote for a President is:
    a. 18-24
    b. 25-44
    c. 45-64
    d. 65 and above

7. The functionalist approach to the question of power in the US is summed up by the term:
    a. cooperation
    b. pluralism
    c. organicism
    d. elitism

8. Mills used the term _____ to refer to those who rule the US.
    a. the ruling class
    b. the power elite
    c. the military-industrial complex
    d. the capitalist class

Ethical Debate Exercise

Divide the class into two groups and have each prepare for 15 minutes, allowing them to prepare for a class debate.

Campaign Finance Reform Debate

Group One:
We should repeal all existing federal limits on how much money individuals or parties can contribute to candidates and instead, establish a campaign finance system that relies solely on disclosure of large contributors. This way the voters can see who is getting what from whom.

Group Two:
We should enforce and strengthen federal limits on campaign contributions. Ban the use of soft money that national party organizations raise on behalf of their candidates.

In Class Activity

Approval voting is a voting system used for elections, in which each voter can vote for as many or as few candidates as the voter chooses. It is typically used for single-winner elections, but can be extended to multiple winners.

Each voter may vote for as many options as he or she chooses, at most once per option. This is equivalent to saying that each voter may "approve" or "disapprove" each option by voting or not voting for it, and it's also equivalent to voting +1 or 0 in a range voting system.

The votes for each option are tallied. The option with the most votes wins.

Approval ballots can be of at least four semi-distinct forms.

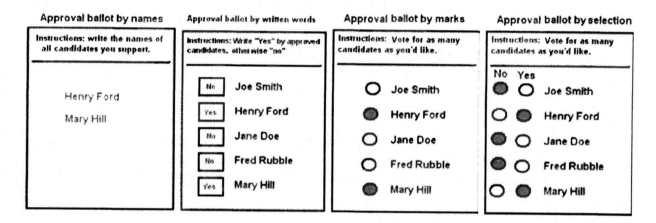

Approval voting Exercise:

Horseback Riding is a growing sport at the 4R Ranch. The Ranch is sponsoring a weekend event and each participant will receive a cowboy hat. Club members decided to let the participants vote on the color of the cowboy hat using Approval Voting. The possible colors are Fuchsia, Purple, Eggshell, Watermelon, and Teal.

> Here is a summary of the results:
>
> **12 participants voted for Fuchsia**
> **7 participants voted for Fuchsia and Teal**
> **20 participants voted for Eggshell and Watermelon**
> **18 participants voted for Purple, Eggshell, and Watermelon**
> **23 participants voted for Teal and Purple**
> **25 participants voted for Watermelon**

Use Approval voting to determine the color of the cowboy hat!

Outside Activity

Research the World of Opportunity available to those wanting to become active in politics!

The Youth Vote Coalition is a national nonpartisan coalition of diverse organizations dedicated to engaging youth between the ages of 18-30 in the political process. The Youth Vote Coalition has over 100 national members who represent young people across the country.
http://www.youthvote.org/

Research shows that the number one reason that young people will vote is if they are asked, so please pledge today to ask someone under thirty you know to vote November 2!

Youth Vote Coalition has organized 20 local Youth Vote Coalitions around the country in 2004. These coalitions help local organizations that promote voter registration, education and get out the vote (GOTV) activities reach out to young people. Local coalitions also assist youth serving organizations or individuals connect with local voting resources to aid young people to participate fully as citizens. When young people are not participating, their valuable voices and ideas are not heard in public policy debates.

Group Project Topics

The Electoral Vote

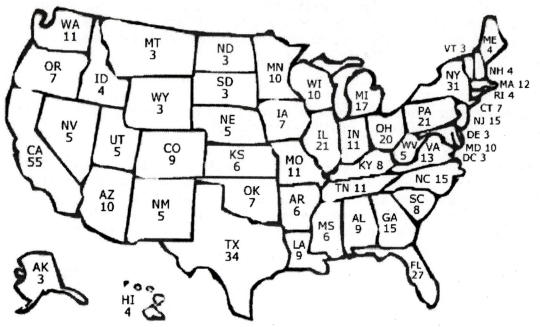

1) Research the history of the Electoral College and all procedures for electing a President.

2) How are electors chosen? Who are voters actually voting for?

3) Why have the electoral college at all? What would it take to get rid of the electoral college?

4) Is it possible for Congress to keep the electoral college and make it fairer?

Ideas for Outside Reading

Basirico, Laurence, A. 1993. "Sociology in Action: Activities for Students." *Harper Collins College Publishers.* New York, NY: Harper Collins.

Brym, Robert J. and Lie, John. 2007. "Sociology." Thomson Wadsworth. Belmont, CA: Wadsworth.

Fowler, Lori, A. 2007. "Introduction to Sociology: A Real World Approach." *Wadsworth.* Belmont, Ca: Wadsworth, Inc.

Osborne, Richard. 2002. "Introducing Sociology." *Icon Books, Ltd.* UK, Europe: Macmillan Distribution Ltd.

Ruane, Janet, M. 1997. "Seeing Conventional Wisdom through the Sociological Eye." *Pine Forge Press.* Thousand Oaks, Ca: Sage.

Westen, Tracy. 1998. "Can Technology Save Democracy?" *National Civic Review.* 26: 47-56.

CHAPTER FIFTEEN – FAMILIES AND INTIMATE RELATIONSHIPS
Video Clip Exercise

Watch the recommended video segments and then answer the questions below.

Video Segment 1:
Wadsworth's Lecture Launchers for Introductory Sociology:
"Family Structures"

Video Clip Exercise:
1. What constitutes a blended family?

2. What are the major roles in the family and how are they defined?

3. How can children benefit from life in an extended family?

Video Segment 2:
Wadsworth's Sociology: Core Concepts
"Family"

Video Clip Exercise:
1. What is a nuclear family and how has it changed?

2. Define an intergenerational family.

3. In what ways are kinship groups formed?

Case Study Exercise

Children of Divorce

A million new children a year are affected by divorce, and many of them may not get the everyday support they need because their parents are preoccupied with this major life crisis. What's more, after a separation or divorce, the oldest child, particularly if it's a girl, will often switch roles, becoming the mother's caretaker and confidante.

Blaire, a 10-year-old fifth grader, had been a fine student and a cheerful child, but all that changed rather suddenly when her parents separated about a month ago. When her dad moved out, Blaire began to miss school quite often. And now, even when she is in the classroom, her thoughts are obviously far away. Her teacher reports that her schoolwork is suffering, but of even greater concern is her loss of interest in her friends. Blaire had been quite the social butterfly, but now she rushes home after school instead of lingering in the school yard to chat or play.

Her teacher said that the situation came to a head recently when Blaire returned from several days' absence looking so sad and tired that she (the teacher) wondered if she was ill. The teacher then decided to call her mother, who, surprisingly, was quite eager to come in and talk. The teacher learned that Blaire has been reluctant to leave her mom since the separation. It was a terrible blow from which the mother has not yet recovered. Blaire often stays home to "take care of" her mom — who apparently confides in her daughter — and to help care for the younger children.

Case Study Exercise:

Is there anything Blaire's teacher can do to help her?

Group Quiz Exercise

Team up in groups to see who can answer the most correctly!

1. The nuclear family is composed of grandparents, aunts, uncles, and cousins.
    a) True
    b) False

2. The family of orientation is formed when a couple has their first child.
    a) True
    b) False

3. Romantic love is composed of an emotional element and a cognitive element.
    a) True
    b) False

4. Exogamy refers to "like marrying like," or marrying within one's own group.
    a) True
    b) False

5. Cohabitation involves adults living together in a sexual relationship without being married.
    a) True
    b) False

6. The family of orientation is the one in which an individual is raised.
    a) True
    b) False

7. Today, the average bride is younger than at any other time in history.
    a) True
    b) False

8. According to Hoschild, all except one of the following are "strategies of resistance" used by men to get out of sharing housework equally with their wives. Which one should not be included?
    a) substitute offerings
    b) waiting it out
    c) appropriate denial
    d) playing dumb

9. Researchers have discovered that romantic love:
    a) first appeared in Medieval Europe
    b) emerged in full form in 1930s Hollywood
    c) has been present in India for 3000 years
    d) was present in a high percentage of early society

10. The tendency of people with similar characteristics to marry one another is known as:
    a) endogamy
    b) homogamy
    c) entrogamy
    d) propinquity

Ethical Debate Exercise

Divide the class into two groups and have each prepare for 15 minutes, allowing them to prepare for a class debate.

The Abortion Issue

Group One:
Life begins at conception. Abortion destroys human life and is morally indefensible. Adoption is a better route.

Group Two:
Every woman has the right to choose what happens to her own body and bearing an unwanted child can harm not only a woman's career, but the child too.

In Class Activity

Child Abuse and Neglect ~ You Decide

Read each of the following situations and decide whether the parents are guilty of neglect, endangerment, physical abuse, or sexual abuse. If the incident is a result of an accident, place an "X" in the blank.

N = Neglect	E = Endangerment	P = Physical Abuse
	S = Sexual Abuse	X = Accident

_____ Amy's mother is in the kitchen fixing dinner, Amy is 3 years old and beginning to explore her surroundings. She sees something up on a shelf and decides she wants it. She starts to climb on the living room furniture while her mother is in the kitchen. She tries to stand on a chair, loses her balance, and hits her head on the side of a table.

_____ Marcus, an 8 year old, has left his home with permission to go to the nearby park. He is climbing on the jungle gym at a playground. He falls and cuts his head. He has a slight concussion and requires stitches.

_____ Jonathan and his friend are playing indoors because it is raining outside. After watching a TV show, they decide to re-enact what they have seen. They dress up and decide to use the real gun in the closet to add to the drama of their Western. As they play, the gun goes off and hits Jonathan in the arm.

_____ Rebecca is taking care of her nieces and nephews while her sister runs out to buy groceries. It is hot outside, so Rebecca decides to let the kids play in the wading pool in the backyard. She tells them to wait for her while she runs upstairs to get some towels. The kids go outside without her, and one of the children slips into the pool face down. Rebecca tries to revive the child, but it is too late. The child dies.

_____ Adam is riding his bike in front of the house. He knows it isn't safe to cross the street, but he figures that his mother will never find out. While crossing the street on his bike, a truck hits him and he suffers serious injuries.

_____ Alisha is sixteen years old and four months pregnant. She uses drugs and alcohol. Her friends have told her that they are worried about both Alisha and the baby.

_____ Carrie lives with her dad who happens to be an alcoholic. He is unemployed and drinks all the time. Carrie is responsible for all of the housekeeping activities. After preparing dinner late one evening, Carrie's father hits her.

_____ Robin is home with her two children. Her 2 year old child is taking a nap. Robin decides to give her infant a bath. She has her in the bathtub when she hears her phone ring in the adjacent room. While on the phone, her infant drowns in the tub.

_____ Patricia is a single mother of two children under the age of five. She is a drug addict and is having difficulty coping. The house is filthy, there is little food, and the children are often left to tend to themselves. The neighbors become concerned and call the police.

_____ Yvonne lives with her mother and her mother's boyfriend. They are often drinking and drugging. Yvonne began to do drugs with her mother, missing school, and separating from friends. A social worker drops in on them one afternoon to find both on drugs. Yvonne and her mother state that the social worker has no right to come into their home, and that Yvonne doesn't have to attend school if she doesn't want to.

_____ Jane and her baby are visiting her baby's father at his home. The cops have had the home under surveillance, and burst in with a search warrant. They take the baby.

Outside Activity

Watch the film, "Tortilla Soup."
Answer the following questions

1) What were some of the sex stereotypes displayed in the film?

2) Did Martin expect his girls to accomplish their gender in any particular way? Cite examples.

3) Was the Naranjo family a nuclear family, or an extended family? Why?

4) Describe the romantic love portrayed in the film, and its two components.

5) Did birth order make a difference in this film? Did the portrayal in the movie match lecture notes?
   a. First born:
   b. Middle born:
   c. Last born:

6) Which of the following trends in United States families were displayed in this film? Cite examples of those that apply.
   a. Postponing marriage:
   b. Cohabitation:
   c. Unwed mothers:
   d. Grand parenting:
   e. The sandwich generation:
   f. Commuter marriages:

Group Project Topics

Social Policies and the Family

Social policies concerning abortion, child care, taxes, maternity and paternity leave, sexual practices, marital trends, custody battles, and divorce rates are repeatedly the subject of political debate.

Have each group member select one social policy and conduct the following research:

1) Look through issues of local newspapers and online libraries to identify social policy controversies on your topic over the last 60 days. Create a reference list that is complete.

    a) Introduce the policy debate and discuss the various positions.
    b) Discuss three reasons to support the policy.
    c) Discuss three reasons to oppose the policy.
    d) Incorporate ideas from the Introductory text to take your own position for or against the policy.

Ideas for Outside Reading

Basirico, Laurence, A. 1993. "Sociology in Action: Activities for Students." *Harper Collins College Publishers.* New York, NY: Harper Collins.

Brym, Robert J. and Lie, John. 2007. "Sociology." Thomson Wadsworth. Belmont, CA: Wadsworth.

Fowler, Lori, A. 2007. "Introduction to Sociology: A Real World Approach." *Wadsworth.* Belmont, Ca: Wadsworth, Inc.

Osborne, Richard. 2002. "Introducing Sociology." *Icon Books, Ltd.* UK, Europe: Macmillan Distribution Ltd.

Rollins, Boyd, C. 1974. "Marital Satisfaction over the Family Life Cycle." *Journal of Marriage and the Family.* 36:271-284.

Ruane, Janet, M. 1997. "Seeing Conventional Wisdom through the Sociological Eye." *Pine Forge Press.* Thousand Oaks, Ca: Sage.

CHAPTER SIXTEEN – EDUCATION
Video Clip Exercise

Watch the recommended video segments and then answer the questions below.

Video Segment 1:
Wadsworth's Lecture Launchers for Introductory Sociology:
"Roles in Education"

Video Clip Exercise:
1. What are the roles defined for students?

2. What are the roles defined for teachers in an academic setting?

3. Does the academic setting change the role for students and teachers?

Case Study Exercise

Watch language grow in the 'Baby Brother' house

http://www.newscientisttech.com/article/dn9167

A baby is to be monitored by a network of microphones and video cameras for 14 hours a day, 365 days a year, in an effort to unravel the seemingly miraculous process by which children acquire language.

Case Study Exercise:

1) In this unusual project, who were the first experimental subjects used in the study?

2) What is this scientist hoping to better understand?

3) Why is the development of language debated among scientists?

4) What is "speechome?" Where is the footage viewed?

5) If successful, what might this experiment help in the future?

Group Quiz Exercise

Team up in groups to see who can answer the most correctly!

1. Which of the following terms represents ways in which a teacher evaluates non–academic behaviors?
   a. Norm–evaluation
   b. Ethnomethodology
   c. Hidden curriculum
   d. Valuatory judgment

2. Which of the following is NOT one of the reasons credentials are required in industrial societies?
   a. urbanization
   b. extremely large size
   c. geographic mobility
   d. anonymity

3. In Japan, competition is:
   a. evident in college admission procedures
   b. discouraged throughout the school years
   c. encouraged beyond elementary years
   d. constantly stressed

4. _____ has been a latent function in US schools, but has now become a manifest function.
   a. transmitting values
   b. child care
   c. teaching patriotism
   d. tracking

5. Tracking is essential for:
   a. social integration
   b. maintaining cultural patterns
   c. gatekeeping
   d. civic socialization

6. When Jennifer's teacher told her what a good student she was for picking up all her crayons, she had experienced:
   a. social modeling
   b. gender clarification
   c. the hidden curriculum
   d. tracking

7. Which of the following is NOT an educational concern among conflict theorists?
   a. reproducing the social class structure
   b. cresting self-fulfilling prophecies
   c. biased IQ tests
   d. schools that are unequally funded

8. All except one of the following are ways in which experts have attempted to raise SAT scores, Which one should NOT be included?

     a. dropping sections of the verbal exam
     b. offering online tutoring
     c. shortening the exam
     d. giving students more time to take the exam

Ethical Debate Exercise

Divide the class into two groups and have each prepare for 15 minutes, allowing them to prepare for a class debate.

IQ Testing

The controversy over intelligence quotient (IQ) tests (also called cognitive ability tests), what they measure, and what this means for society has not abated since their initial development. IQ tests rely largely upon Symbolic Logic as a means to scoring, and as Symbolic Logic is not inherently synonymous with intelligence, the question remains as to exactly what is being measured via such tests.

Group One:
IQ scores demonstrate that power and wealth will always be distributed unequally. IQ tests should be maintained and used to weed out those of higher ability from those of lesser ability.

Group Two:
IQ tests do not measure intelligence, but rather a specific skill set valued by those who create IQ tests. They should not be used as sole measures of acceptance.

In Class Activity

The Hidden Curriculum
Group Activity

1) You've just received word that you'll be substituting today in a 3rd grade classroom. You notice two children. One seems happy, adjusted, and outgoing. The other seems quiet, reclusive, and depressed. What could you do or say, specifically, in the span of 4 hours that could change their lives for the better?

2) A studious child comes to the front of the class to ask you a question. As she does, several students behind her start snickering about "Goodie-Two-Shoes." What can you do in front of the class to change that behavior?

3) A parent of one of your students walks in the class unannounced. He/she happens to be a very close friend of yours from years back. The two of you start talking about "old times." You make plans, in front of the children, to get together for dinner soon. You notice the child of that parent looks mortified. The other students begin teasing that child about "dinner with the teach." Damage done. What can you do to immediately resolve the issue?

4) Two of your students are called out of class and asked to go meet with the principal. Rumors begin flying, and other students are naturally curious. How do you stop the talk and protect their privacy at the same time?

5) A student approaches you during an exam to ask a question. You "lead" him/her to derive at the correct answer on his/her own. The other students feel you "gave him/her the correct answer." What can you do?

6) You are now teaching a junior high school class. A female student has been asked out by a male classmate. She declines the offer. He then harasses her about the rejection. She approaches you to let you know she is now uncomfortable. You, as the teacher, are liable for the atmosphere in the classroom at all times. You have been notified that she is uncomfortable. What do you do?

Outside Activity

Report Card Comments

View the following <u>real examples</u> of written comments to be used on children's report cards. Decide whether you think the comment is useful, harmful, or neutral.

If harmful, correct the comment into a useful phrase.
Place a "U" next to the useful comments.
Place an "N" next to the neutral comments.
Place an "H" next to the harmful comments. Correct all harmful comments.

1.   Is learning to share and listen.
2.   Is developing a better attitude toward ___ grade.
3.   Is showing interest and enthusiasm for the things we do.
4.   Is learning to occupy his time constructively.
5.   Can be very helpful and dependable in the classroom.
6.   Has great potential and works toward achieving it.
7.   Is learning to be a better listener.
8.   Is learning to be careful, cooperative, and fair.
9.   Is continuing to grow in independence.
10. Enthusiastic about participating.
11. Gaining more self-confidence.
12. Has a pleasant personality.
13. Is learning to listen to directions more carefully.
14. Now accepts responsibility well.
15. _____'s work habits are improving.
16. Has shown a good attitude about trying to improve in ___.
17. Seems eager to improve.
18. Is cooperative and happy.
19. Is willing to take part in all classroom activities.
20. Works well with her neighbors.
21. _____'s attitude toward school is excellent.
22. Has the ability to follow directions.
23. Has a sense of humor and enjoys the stories we read.
24. Is very helpful about clean-up work around the room.
25. Anxious to please.
26. Needs to increase speed and comprehension in reading.
27. Gets along well with other children.
28. Your constant cooperation and help are appreciated.
29. Works well in groups, planning and carrying out activities.
30. Seems to be more aware of activities in the classroom.
31. Takes an active part in discussions pertinent to ___.
32. Extremely conscientious.
33. Needs to develop a better sense of responsibility.
34. Performs well in everything he undertakes.
35. Unusually mature.
36. Rate of achievement makes it difficult for ___ to keep up with the class.

37. _____'s academic success leaves much to be desired.
38. Cries easily.
39. Needs to listen to directions.
40. Needs to be urged.
41. Uses colorful words
42. Reverses some numbers still
43. Is too easily distracted
44. Does colorful and interesting art work
45. Requires too much supervision.
46. Please encourage him to do things on his own.
47. Thanks for the help I know you have given her.
48. _____ would benefit from reading many library books this summer.

Group Project Topics

Praise versus Encouragement

Most of us believe that we need to praise our children more. However, there is some controversy regarding this point. If we always reward a child with praise after a task is completed, then the child comes to expect it. Children who are subjected to endless commentary, acknowledgment, and praise eventually learn to do things not for their own sake, but to please others.

Praise	Encouragement
Your are the best student I ever had.	You are a fine student. Any teacher will appreciate and enjoy you.
You are always on time.	You sure make an effort to be on time.
You have the highest score in the class on this exam.	You did very well on this exam.
I am so proud of you.	You seem to really enjoy learning
You're the best helper I ever had.	The room looks very neat since you straightened the bookshelves.
I'm so proud of your artwork.	It is nice to see that you enjoy art.

Determine whether each of the following statements is one of praise or encouragement. Place a P or E in front of each statement. If the statement is a Praise statement, change it into an Encouragement statement.

_____ I like the way Brian is sitting with his legs crossed.

_____ What a good job of putting away the blocks, Tyrone.

_____ I noticed that you have used red paint and green paint at the top of the picture.

_____ Billy, what a beautiful painting.

_____ You helped us set the table.

_____ How did you build your animal hospital, Ellie?

_____ I like the way you printed your name.

_____ Very good Pam, you walked into the classroom.

_____ Great job, Ofelia.

Have students in the group research the difference between praise and encouragement. Answer the following below, and then design a ten minute classroom lecture using the appropriate methods on the subject of your choice.

Ideas for Outside Reading

Bolton, Robert (1979) " People Skills," NY:Simon & Shuster, pg. 43-44, 181.

Brym, Robert J. and Lie, John. 2007. "Sociology." Thomson Wadsworth. Belmont, CA: Wadsworth.

Grunwald, B. B. and Pepper, F. C. (1982). "Maintaining Sanity in the Classroom, Classroom Management Techniques, 2nd Ed." NY:HarperCollins, pg. 110.

Rosenberg, Marshall B. (2005) *Speak Peace in a World of Conflict, What you say next will change your world*, PuddleDancer Press: Encinitas, CA, Chapter 12: Gratitude.

CHAPTER SEVENTEEN – RELIGION
Video Clip Exercise

Watch the recommended video segments and then answer the questions below.

Video Segment 1:
Wadsworth's Lecture Launchers for Introductory Sociology:
"The Role of Religion"

Video Clip Exercise:
1. If religion creates a moral community in a culture, is there room for differing religions in one culture?

2. How has the role of women in religion changed?

3. How has the function of religion changed?

Video Segment 2:
Wadsworth's Sociology: Core Concepts:
"Religion"

Video Clip Exercise:
1. How does religion serve as a force of social control in any society?

2. Give examples of norms common to most religions.

3. What challenges do religions face?

Case Study Exercise

A Second Alienation?

http://www.jewishworldreview.com/0199/lipschutz1.asp

Long the best fit for alienated outsider status in an overwhelmingly Christian America, the secular Jew has moved from culturally critical minority toward microscopic marginality.

Case Study Exercise:

1) Compare the treatment of Jews in America over time, from post WWII to the present.

2) What kind of alienation is the author speaking of, and why?

3) What is a self-identifying trait for American Jews? Explain.

4) What role does perception play in this sense of alienation among Jewish Americans?

Group Quiz Exercise

Team up in groups to see who can answer the most correctly!

1. Which of the following is NOT a part of Durkheim's definition of religion?
    a. a belief in the supernatural.
    b. a moral community
    c. beliefs
    d. practices

2. In Durkheim's terms, religion uses symbols to separate the:
    a. supernatural from the natural
    b. church from the secular community
    c. sacred from the profane
    d. clergy from the laity

3. Religions use rituals to:
    a. instill fear
    b. instill obedience
    c. unite people into a moral community
    d. hide true meanings

4. When Marx stated that religion was the opium of the people, he meant that:
    a. it becomes an addiction
    b. it obscures the real meaning of materialistic life
    c. it takes the mind of the worker off of the unfairness of inequality
    d. it masks the dangers of class consciousness

5. Modernization signifies:
    a. a unified picture of the world
    b. a religious explanation of life
    c. a sudden awareness of the supernatural
    d. the transformation of traditional societies into industrial ones

6. At present, the Protestant ethic and the spirit of capitalism:
    a. can be seen operating in industrialized nations
    b. are held in check by global capitalism
    c. are no longer operating as Weber predicted
    d. are not confined to any specific religion or part of the World

7. Which of these is NOT characteristic of cults?
    a. they have charismatic leaders
    b. the use thought control
    c. most of them fail
    d. they are at odds with the dominant culture

8. An ecclesia is:
     a. a very large religious body
     b. an old, exhausted denomination
     c. a state religion
     d. an international religious body

Ethical Debate Exercise

Divide the class into two groups and have each prepare for 15 minutes, allowing them to prepare for a class debate.

Prayer in the Classroom

The first American school system was founded in Massachusetts, in 1647, and was established to ensure that children would grow up with the ability to read the Bible.

Group One:
Public schools had prayer for nearly 200 years before the Supreme Court ruled that state-mandated class prayers were unconstitutional. The fact that prayer was practiced for nearly 200 years establishes it by precedent as a valid and beneficial practice in our schools. Our government was based on religious principles from the very beginning.
To forbid the majority the right to pray because the minority object, is to impose the irreligion of the minority on the religious majority. Forbidding prayer in schools, which a three-quarters majority of Americans favors, is the tyranny of the minority. It is minority rule, not democracy.

Group Two:
The First Amendment says "Congress shall make no law respecting an establishment of religion". Just because the First Amendment wasn't adhered to doesn't mean it's not the law of the land or right. The framers of the Constitution knew the evils of state sponsored religion. The very first line in the very first amendment of the Bill of Rights reads "Congress shall make no law respecting an establishment of religion, or prohibiting the free exercise thereof". After these were assured only then was the Constitution ratified.

In Class Activity

Durkheim and Totemism

"A religion is a unified system of beliefs and practices relative to sacred things, that is to say, things set apart and forbidden--beliefs and practices which unite in one single moral community called a Church, all those who adhere to them."

A totem is any natural or supernatural object, being or animal which has personal symbolic meaning to an individual and to whose phenomena and energy one feels closely associated with during one's life.

In modern times, some individuals, not otherwise involved in the practice of a tribal religion, have chosen to adopt as a personal totem an animal which has some kind of special meaning to them.

1) If, According to Durkheim, a totem is any object which has symbolic meaning to an individual, many objects in modern times could become totems. Think of an object in modern times that people have come to worship. Draw it here.

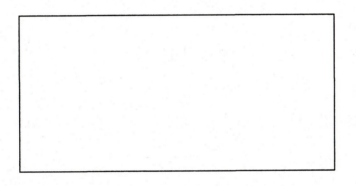

2) Has your modern totem grown to occupy religious significance of any kind? Does it separate the sacred from the profane in any way? Explain your answer using Durkheimian terms.

Outside Activity

Using Census Data to Examine Religious Affiliation

There are a number of variables associated with religious affiliation: income, education, race, ethnicity, and others. Visit the census bureau online in order to answer the following questions.

http://www.census.gov/

1) Does social class have anything to do with religious affiliation?

2) Do race and or ethnicity have anything to do with religious affiliation?

3) Does education have anything to do with religious affiliation?

4) What percentage of those in the lower classes are Catholic? Jewish? Protestant?

5) Are Jewish individuals more apt to vote Democratic? Are Christians more apt to vote Republican?

6) Identify five occupations in which the types of data about religion might be useful, and discuss how. i.e., How might is be useful for a politician to know what religious groups reside in a particular location?

Group Project Topics

The Role of Women in Religion

Have each group member select a different religion and analyze the role of women in that particular religious organization.

Be sure to include:
historical developments
political and legal developments
participation in services
martyrdom

Ideas for Outside Reading

Brym, Robert J. and Lie, John. 2007. "Sociology." Thomson Wadsworth. Belmont, CA: Wadsworth.

Campbell, Anne. 2002. "A Mind of Her Own. The Evolutionary Psychology of Women. "Oxford University, Oxford. p. 91.

Johnson, Buffie. 1988. "Lady of the Beasts." San Francisco: Harper & Row.

Richard J. Lane & Jay Wurtz, In Search of the Woman Warrior, (Element, Boston, 1998) p. 44.

CHAPTER EIGHTEEN – HEALTH, HEALTH CARE, AND DISABILITY
Video Clip Exercise

Watch the recommended video segments and then answer the questions below.

Video Segment 1:
Wadsworth's Lecture Launchers for Introductory Sociology:
"Alternative Health Care Methods"

Video Clip Exercise:
1. What is the focus of Mexican folk medicine?

2. Are alternative methods of healing valuable?

3. Do medical practices from other cultures meet needs that traditional medicine cannot?

Map the Stats Exercise

NUMBER OF AIDS CASES

50 STATES

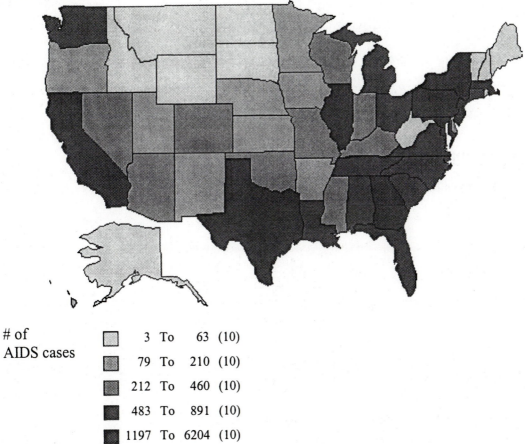

of
AIDS cases

☐	3	To	63	(10)
☐	79	To	210	(10)
☐	212	To	460	(10)
■	483	To	891	(10)
■	1197	To	6204	(10)

Map the Stats Exercise:

(1) Look at this map of the United States and identify those states with the highest number of AIDS cases.

(2) Which states have the lowest number of cases?

Case Study Exercise

The Canadian Health Care System

---

http://www.canadian-healthcare.org/

Canada's health care system is a group of socialized health insurance plans that provides coverage to all Canadian citizens.

Case Study Exercise:

1) Who funds Canada's health care system? How?

2) What kind of care is provided under this health care plan?

3) Why is Canada's health care system the center of such political debate?

4) What is the life expectancy and infant mortality rate in Canada? Why is this important?

Group Quiz Exercise

Team up in groups to see who can answer the most correctly!

1. _____ is a state of complete physical, mental, and social well-being.
   - a. health
   - b. wealth
   - c. popularity
   - c. clarity

2. Most historical preindustrial societies were able to maintain a healthy environment.
   - a. true
   - b. false

3. _____ has dramatically improved human health.
   - a. technology
   - b. medicine
   - c. industrialization
   - d. all of these choices

4. _____ is the study of how health and disease are distributed throughout a society's population.
   - a. demography
   - b. geography
   - c. social epidemiology
   - d. infant mortality

5. Death is now rare among young people, with the exception of mortality resulting from accidents and AIDS.
   - a. true
   - b. false

6. Which of these tops the list of preventable health hazards?
   - a. AIDS
   - b. cigarette smoking
   - c. traffic accidents
   - d. STDs

7. A(n) _____ is an intense involvement in dieting or other forms of weight control in order to become very thin.
   - a. STD
   - b. eating disorder
   - c. psychological disorder
   - d. body dysmorphic disorder

8. _____ is an institution concerned with combating disease.
   - a. Health care
   - b. Demography
   - c. Medicine
   - d. Holistics

Ethical Debate Exercise

Divide the class into two groups and have each prepare for 15 minutes, allowing them to prepare for a class debate.

Should Marijuana be a Medical Option Now?

Group One:
The evidence is overwhelming that marijuana can relieve certain types of pain, nausea, vomiting and other symptoms caused by such illnesses as multiple sclerosis, cancer and AIDS -- or by the harsh drugs sometimes used to treat them. Indeed, marijuana is less toxic than many of the drugs that physicians prescribe every day.

Group Two:
Smoked marijuana damages the brain, heart, lungs, and immune system. It impairs learning and interferes with memory, perception, and judgment. Smoked marijuana contains cancer-causing compounds and has been implicated in a high percentage of automobile crashes and workplace accidents. The major potential pulmonary consequences of habitual marijuana use of particular relevance to patients with AIDS is superimposed pulmonary infection, which could be life threatening in the seriously immonocompromised patient.

In Class Activity

Benefits Who?

Below you will find a list of social problems facing our health care industry today.

For each issue, answer the following:
1) Who is hurt by the problem?
2) Who benefits by the problem?
3) How much money is made by the problem?
4) Who determines possible solutions?

Drug Use:
1-
2-
3-
4-

Cancer:
1-
2-
3-
4-

Eating Disorders:
1-
2-
3-
4-

Teen Pregnancy:
1-
2-
3-
4-

AIDS:
1-
2-
3-
4-

Smoking:
1-
2-
3-
4-

Outside Activity

Disability Awareness

Assign students randomly to wheelchairs, and have them perform the following activities in pairs in order to increase disability awareness. Have the student wear a sign, "Social Experiment" in order to prevent any negative disruptions.

You will be placed in groups of two, you cannot allow your teammates to assist you in any way! You may only ask for assistance from a stranger if absolutely necessary. Your teammates are to be used only for safety purposes. You have 30 minutes to complete the exercise.

Each student must find and bring back the following objects:

1) Go to the disability office on campus and collect a brochure on the disability of your choice.

2) Go to the registrar and use a computer terminal to verify your registration. Print out a copy of your current schedule. If this is not appropriate for your campus, pick up a campus catalog.

3) Go to the library and collect any flier. Have the librarian time stamp or sign the fliar.

4) Make a phone call from any campus phone or pay phone to my office phone at _____ and leave a message with your full name and time of call.

5) Go to the theatre and collect a flier on upcoming dramatic presentations or events.

6) Grab a school newspaper.

Return to class with all items!

Group Project Topics

Current Health Care Issues

Have each group member read about the different health care programs available in the United States. Familiarize yourself with the sociological theoretical approaches as well. Interview a hospital administrator, a physician, nurse, or administrator and record their responses to the following questions.

1- What do you see as the three biggest issues facing our health care system today?
2- What do you feel should be done to help resolve these issues?
3- Should our government play a role in resolving these issues?

Use functionalism, conflict theory, or symbolic interactionism to interpret your interview results.

Ideas for Outside Reading

John Sheils and Randall Haught, "The Cost of Tax-Exempt Health Benefits in 2004," *Health Affairs* Web Exclusive, February 25, 2004, at *http://content.healthaffairs.org/cgi/reprint/hlthaff.w4.106v1* (July 20, 2005).

Robert B. Helms, "The Tax Treatment of Health Insurance: Early History and Evidence, 1940–1970," in Grace-Marie Arnett, ed., *Empowering Health Care Consumers Through Tax Reform* (Ann Arbor: University of Michigan Press, 1999), p. 11.

Carmen DeNavas-Walt, Bernadette D. Proctor, and Robert J. Mills, "Income, Poverty, and Health Insurance Coverage in the United States: 2004," U.S. Census Bureau, August 2005, p. 16.

Sara R. Collins, Karen Davis, and Alice Ho, "A Shared Responsibility: U.S. Employers and the Provision of Health Insurance to Employees," Commonwealth Fund, *In the Literature* Publication No. 839, June 2005.

Kaiser Family Foundation and the Health Research and Educational Trust, "Distribution of Firms Providing a Choice of Health Plans, by Firm Size, 2005," *Employer Health Benefits 2005 Annual Survey*, Exhibit 4.2.

CHAPTER NINETEEN – POPULATION AND URBANIZATION
Video Clip Exercise

Watch the recommended video segments and then answer the questions below.

Video Segment 1:
Wadsworth's Lecture Launchers for Introductory Sociology:
"Urban Issues"

Video Clip Exercise:
1. Explain gentrification and who is affected by it.

2. Who benefits from gentrification?

3. What are the social benefits of green space?

Case Study Exercise

Remember the Bowery: A Case Study in Gentrification
Lessons for New Orleans

http://www.utne.com/webwatch/2005 223/news/11868-1.html

The phrase "rebuilding the Gulf Coast" has buzzed through the media for several weeks. Terms like "rebuilding" and "redevelopment" raise red flags for many activists, though, because they're often code words for gentrification.

Case Study Exercise:

1) What is the Bowery and where is it located?

2) What change has occurred on the "lower east side?"

3) What is one of the hallmarks of gentrification?

4) What is the best cure for gentrification?

Group Quiz Exercise

Team up in groups to see who can answer the most correctly!

1. _____ is the study of human population.
    a. demography
    b. sociology
    c. geography
    d. anthropology

2. Which of the following represents the incidence of childbearing in a society's population?
    a. fertility
    b. mortality
    c. migration
    d. immigration

3. Which of the following represents the movement of people into and out of a specified territory?
    a. fertility
    b. mortality
    c. migration
    d. immigration

4. Which of the following represents the incidence of death in a society's population?
    a. fertility
    b. mortality
    c. migration
    d. immigration

5. _____ is a thesis linking population patterns to a society's level of technological development.
    a. gentrification
    b. Sapir-Whorf hypothesis
    c. Demographic Transition Theory
    d. Malthusian Theory

6. A _____ is a vast urban region containing a number of cities and their surrounding suburbs.
    a. urbanite
    b. suburbs
    c. megalopolis
    d. gesellschaft

7. _____ is the study of the interaction of living organisms and the natural environment.
    a. population
    b. environmentalism
    c. ecology
    d. urbanism

8. Who coined the terms Gesellschaft and Gemeinschaft?
   a. Marx
   b. Durkheim
   c. Weber
   d. Tonnies

Ethical Debate Exercise

Divide the class into two groups and have each prepare for 15 minutes, allowing them to prepare for a class debate.

The Malthusian Debate

It is currently estimated that there is, or there will be shortly, six billion humans inhabiting the planet earth. The theme of population, and more specifically, overpopulation has been in the popular mind for the last thirty years or more.

Group One:
Rapid population growth represents a major obstacle to development, as valuable resources are diverted from productive expenditure to the feeding of a growing population. Population growth has a negative and potentially destructive impact on the environment. Even if a growing population can be fed, the environment cannot sustain such large numbers, population growth will lead to the explosion of pollution, which will have a catastrophic effect on the environment.

Group Two:
The growth of population has the potential to stimulate economic growth and innovation. More people means more problem solvers, since human creativity has the potential to overcome the limits of nature. The market mechanism can help establish a dynamic equilibrium between population growth and resources.

In Class Activity

Infant Mortality Exercise

See the top 10 countries with the highest infant mortality rates below. Plot them on a world map.

<u>Country</u>	<u>Value</u> / <u>Unit</u>	
Angola	192.50	deaths/1,000 live births
Afghanistan	165.96	deaths/1,000 live births
Sierra Leone	145.24	deaths/1,000 live births
Mozambique	137.08	deaths/1,000 live births
Liberia	130.51	deaths/1,000 live births
Niger	122.66	deaths/1,000 live births
Somalia	118.52	deaths/1,000 live births
Mali	117.99	deaths/1,000 live births
Tajikistan	112.10	deaths/1,000 live births
Guinea-Bissau	108.72	deaths/1,000 live births

http://www.geographyiq.com/ranking/ranking_Infant_Mortality_Rate_top25.htm

Outside Activity

How Fertility, Mortality, and Migration Affect Population

These factors can have significant impacts on the welfare and types of issues countries may face. Read the chapter in your introductory text on population and urbanization in order to familiarize yourself with these issues.

1) Define fertility, mortality, and migration in detail.

2) Look through newspapers and magazines in order to identify articles on:

Starvation
Overpopulation
The spread of disease
Unemployment

3) Describe how the United States intends to deal with the above social ills. What policies have been put in place to prevent or cure such problems?

4) How does fertility play a role in your research findings?

5) How does mortality play a role in your research findings?

6) How does migration play a role in your research findings?

Group Project Topics

Population Facts - the Future in just 22 years

Scarce Water --- Currently, 434 million people face either water stress or scarcity.

Scarce Cropland ---The number of people living in countries where cultivated land is critically scarce is projected to increase to between 600 million and 986 million in **2025**.

Forests --- Based on the medium population projection and current deforestation trends, by **2025** the number of people living in forest-scarce countries could nearly double to 3 billion.

Species Extinction --- More than 1.1 billion people live in areas that conservationists consider the most rich in non-human species and the most threatened by human activities.

Is global warming a real threat, or a political tactic? Is the data listed above factual? Divide into teams to conduct research on population growth and environmental strain. Present your findings to the class.

Ideas for Outside Reading

Basirico, Laurence, A. 1993. "Sociology in Action: Activities for Students." *Harper Collins College Publishers.* New York, NY: Harper Collins.

Brym, Robert J. and Lie, John. 2007. "Sociology." Thomson Wadsworth. Belmont, CA: Wadsworth.

Fowler, Lori, A. 2007. "Introduction to Sociology: A Real World Approach." *Wadsworth.* Belmont, Ca: Wadsworth, Inc.

Osborne, Richard. 2002. "Introducing Sociology." *Icon Books, Ltd.* UK, Europe: Macmillan Distribution Ltd.

Ruane, Janet, M. 1997. "Seeing Conventional Wisdom through the Sociological Eye." *Pine Forge Press.* Thousand Oaks, Ca: Sage.

CHAPTER TWENTY – COLLECTIVE BEHAVIOR, SOCIAL MOVEMENTS, AND SOCIAL CHANGE
Video Clip Exercise

Watch the recommended video segments and then answer the questions below.

Video Segment 1:
Wadsworth's Lecture Launchers for Introductory Sociology:
"AIDS Action: One Man's Personal Story"

Video Clip Exercise:
1. What does it mean to translate a personal problem into a social issue?

2. How does Victor Ayala's research and action provide a positive model for others?

3. Describe the empirical research of Victor Ayala and the variables he examined.

Video Segment 2:
Wadsworth's Lecture Launchers for Introductory Sociology:
"Cathedrals of Consumption"

Video Clip Exercise:
1. What are some examples of "cathedrals of consumption?"

2. Beyond consumption, what other functions do these structures provide in our culture?

3. How have the "cathedrals of consumption" changed our society?

Case Study Exercise

Gang Membership from a Sociological Perspective

Street gangs have a presence in 94 percent of all U.S. cities with populations greater than 100,000.
Los Angeles has over 950 different gangs with more than 100,000 members.
One study performed in Chicago showed that 5 percent of elementary school children were affiliated with street gangs, as were 35 percent of high school dropouts.
Membership in a street gang increases one's risk of violent death by 60 percent.

Case Study Exercise:

1) Why do you feel gang membership is so prevalent today? Why are youth so attracted to this?

2) Are you surprised that elementary school children are affiliated with gangs at such a young age? Why or why not?

3) Why do you believe that membership in a street gang increases one's risk of violent death? Aren't they supposed to protect one another?

4) How would you analyze gang membership from a sociological point of view?

Group Quiz Exercise

Team up in groups to see who can answer the most correctly!

1. _____ is defined as activity involving a large number of people, often spontaneous and controversial.
   a. crowd behavior
   b. mob behavior
   c. collective behavior
   d. riot behavior

2. Why is the study of collective behavior difficult?
   a. it covers a wide variety of behavior
   b. it is complex
   c. much of the behavior is transitory
   d. all of these choices

3. _____ are temporary gatherings of people who share a common focus.
   a. gangs
   b. cliques
   c. mobs
   d. crowds

4. A _____ is a social eruption that is highly emotional, violent, and undirected.
   a. riot
   b. revolt
   c. dispute
   d. debate

5. Ordinary people typically gain power only by acting collectively.
   a. true
   b. false

6. Because crowds have been able to affect social change, they pose a threat to elitist power.
   a. true
   b. false

7. _____ behavior refers to collective behavior among people dispersed over a wide geographical area.
   a. riot
   b. contagion
   c. mass
   d. panic

8. _____ is unsubstantiated information spread informally.
   a. gossip
   b. rumor
   c. untruth
   d. an urban legend

Ethical Debate Exercise

Divide the class into two groups and have each prepare for 15 minutes, allowing them to prepare for a class debate.

Government Surveillance of Social Movements

A public policy debate has emerged over whether the US government should have a free hand to spy on the activities of social movements.

Group One:
We need government regulations requiring that all encryption programs be built with a "backdoor," or an encryption key that would allow officials to read coded messages quickly and easily.

Group Two:
The government should NOT be given more power to invade people's privacy. The government has a history of surveillance of popular social movements like the Civil Rights Movement.

In Class Activity

Crowd Behavior

A crowd is a collection of people temporarily brought together for some common activity. While some types of crowds are spontaneous, others are the result of much forethought.

Look through a newspaper or magazine and find one article which depicts a crowd of at least 4 or more people engaged in any activity. Use your text to answer the following questions:

1- Which type of crowd is exhibited in your article?

2- Why do you believe this crowd has congregated? Did they spontaneously come together, or did this event require much forethought?

3- Is this a stable crowd, or an unstable one? Do you feel they will come together again?

4- Are the group members proud of their group membership, or ashamed?

5- What is the ultimate goal of this gathering? How will the crowd ultimately breakdown?

Outside Activity

Censorship in the United States

Although freedom of the press and speech are guaranteed to Americans under the 1ˢᵗ Amendment, censorship remains a controversial issue alive and well within our borders.

1- Look through internet sources to identify two censorship cases that have occurred within the last two months (flag burning, music censorship, art censorship, etc.) Attach the articles to this worksheet.

2- Discuss which groups are censoring the items, and why you think they are doing so. What are the functions and dysfunctions of censorship? What are the benefits to each of the groups if they maintain their positions?

3- Are the censors acting on a collective agenda, or a personal one? Who are the competing groups involved?

4- State whether you are for or against censorship in relation to your two cases. Why?

Group Project Topics

The Impact of Strikes

The most important strike in American labor history, historians agree, began at the end of 1936. The feisty young United Auto Workers launched the first of a series of sit-down strikes against General Motors at Fisher Body Plant No. 1 in Flint.

Research the History of Strikes in the United States, beginning with the Taft-Hartley Act. Have each group member select a different organization and analyze its history and development through history to the present.

Each group member should analyze the following for their organization:
1- Record the History of the Strike Movement within Your Organization
2- Record the success and losses of the strikes
3- Record whether or not it was beneficial to the organization to strike
4- Incorporate Resource Mobilization Theory into your analysis

Ideas for Outside Reading

Basirico, Laurence, A. 1993. "Sociology in Action: Activities for Students." *Harper Collins College Publishers.* New York, NY: Harper Collins.

Brym, Robert J. and Lie, John. 2007. "Sociology." Thomson Wadsworth. Belmont, CA: Wadsworth.

Demarest, David P., ed. "The River Ran Red": Homestead 1892. Pittsburgh, University of Pittsburgh Press, 1992.

Montgomery, David. The Fall of the House of Labor: The Workplace, the State, and American Labor Activism, 1865-1925. New York: Cambridge University Press, 1987.

Smith, Carl. Urban Disorder and the Shape of Belief. Chicago: University of Chicago Press, 1995.

Answer Key Chapter One

Map the Stats Exercise:

(1) Suicide is least likely to occur in California, Minnesota, Illinois, Ohio, New York, Maryland, New Jersey, Connecticut, and Massachusetts. Suicide is most likely to occur in Washington, Oregon, Idaho, Montana, Wyoming, Nevada, Colorado, Arizona, New Mexico, Oklahoma, and Alaska.

(2) Rain; Isolation; A greater sense of community or contact with others and different weather patterns.

Group Quiz Exercise:

1. A
2. D
3. C
4. A
5. B
6. B – there are numerous theories to choose from.
7. A
8. B – when relatives are home on a Sunday.
9. A
10. A

Outside Activity:

1. **Sociological perspective** is the most general paradigm (a point of view, a distinct way of thinking) specific to the field of sociology. Sociological perspective focuses not on individuals but their group, or society, and attempts to explain human social structures, including cultural and governmental institutions and forms of activity and interpersonal relations using social facts or social forces.

2. **Sociological imagination** is a sociological term coined by American sociologist C. Wright Mills in 1959. The ability to connect seemingly impersonal and remote historical forces to the most basic incidents of an individual's life. It suggests that people look at their own personal problems as social issues and, in general, try to connect their own individual experiences with the workings of society. The sociological imagination enables people to distinguish between personal troubles and public issues.

3. Yes, they are really the same thing.

4. No - http://www.sociology.org.uk/p2i6.htm

Answer Key Chapter Two

Map the Stats Exercise:

(1) Europe, Russia, and Asia have the highest number of physicians. Africa and Papua New Guinea have the least amount of physicians. It is important to compare the United States with Canada and South America because they have similar amounts of physicians although they have entirely different health care systems and standards of living.

(2) This is significant because of the resources available to lend solutions to this problem in modern times. AIDS hinders doctors in Europe from moving into Africa to lend aid. South America and Asia share similar physician rates.

Group Quiz Exercise:

1. B
2. A
3. D
4. A
5. A
6. D
7. D
8. C
9. B
10. B

Answer Key Chapter Three

Map the Stats Exercise:

(1) The majority of female headed households living below the poverty line are located in the South. California and Nevada seem to be surrounded by higher levels of poverty stricken households.

(2) The southern region of the United States is so inundated with poverty because of different minimum wage pay rates, different poverty line index payouts, and lower standards of living.

Case Study Exercise:

(1) monuments

(2) story; stories

(3) resist

Group Quiz Exercise:

1. B
2. A
3. C
4. A
5. B
6. B
7. C

Answer Key Chapter Four

Map the Stats Exercise:

(1) The word obedient has different meaning and interpretations in different cultures around the globe. This needs to be taken into consideration when examining data.

(2) Sweden and Germany are least likely to think that is important for children to be obedient? This does not mean these countries wish for their children to be lawless.
36.8 to 45.8% think it very important that a child be obedient in the United States.

Case Study Exercise:

(1) kind of; whom they watch them with

(2) His main job is to perform research exploring the ways media shape children's development.

(3) violent; less time

(4) Studies have found that the average school-age child spends 27 hours a week watching TV.

(5) Opinion answer

Group Quiz Exercise:

1. D
2. C
3. D
4. C
5. D
6. B
7. A
8. A
9. A
10. B

| Answer Key Chapter Five |

Case Study Exercise:

(1) Chronic liars are capable, successful, even disciplined people who embellish their life stories needlessly. They don't suffer from an established mental illness, as many habitual fabricators do. They're just...liars. These men and women are viewed as otherwise normal.

(2) 5

(3) The average fib rate: three for every 10 minutes of conversation.

(4) To avoid hurting other people's feelings, to cover our own embarrassment, to reassure the needlessly anxious, to spare unnecessary headaches.

(5) On psychological tests, chronic liars do show evidence of a neurological imbalance. They have highly developed verbal skills combined with slight impairment in the frontal lobes of the brain.

Group Quiz Exercise:

1. C
2. B
3. B
4. A
5. D
6. B
7. C
8. B
9. D
10. A

Answer Key Chapter Six

Group Quiz Exercise:

1. B
2. A
3. A
4. B
5. B
6. C
7. A
8. C
9. C
10. A

Outside Activity Exercise:

(1) Corporate culture, like any other culture, is a set of behaviors and codes that guide interaction. Corporate culture is expressed in the architecture and interior design of offices, by what people wear to work, by how people communicate with each other, and in employee titles.

(2) An organization's culture is not necessarily the company's mission and list of values as spelled out on a plaque in the lobby. These are ideals, which may or may not be manifest in the corporate culture.

(3) The culture of an organization operates at both conscious and unconscious levels. At its most basic, it involves deeply rooted beliefs, values, and norms held by members of the organization. Insiders may find it difficult or even impossible to recognize these deeply held assumptions, particularly if they have "grown up" in the organizational culture. Espoused or secondary values, on the other hand, exist at a more conscious level; these are the values people discuss, promote, and try to live by.

Answer Key Chapter Seven

Map the Stat Exercise:

(1) Russia has the highest acceptance of euthanasia. South America, India, and South Africa have the least acceptance of euthanasia.

(2) Increasing older populations put financial and social pressures on government services and allocation of funds.

Group Quiz Exercise:

1. C
2. B
3. C
4. A
5. C
6. B
7. A
8. B
9. C
10. A

Outside Activity Exercise:

b. The neighbor who invited Megan Kanka to see his puppy was a twice-convicted pedophile, who raped and murdered her, then dumped her body in a nearby park.

b. The statutes require states to establish registration programs so local law enforcement will know the whereabouts of sex offenders released into their jurisdictions, and notification programs so the public can be warned about sex offenders living in the community.

Answer Key Chapter Eight

Case Study Exercise:

(1) Three billion people live on less than two dollars a day.

(2) The GDP (Gross Domestic Product) of the poorest 48 nations (i.e. a quarter of the world's countries) is less than the wealth of the world's three richest people combined.

(3) Nearly a billion people entered the 21st century unable to read a book or sign their names.

(4) The wealthiest nation on Earth has the widest gap between rich and poor of any industrialized nation.

(5) The lives of 1.7 million children will be needlessly lost this year because world governments have failed to reduce poverty levels.

(6) According to UNICEF, 30,000 children die each day due to poverty.

Group Quiz Exercise:

1. C
2. D
3. B
4. B
5. A
6. B
7. A
8. A
9. B
10. A

In Class Activity:

(1) The United States shares similar poverty levels with Canada.
(2) China shares similar poverty levels with Turkey.
(3) Mexico shares similar poverty levels with Ecuador and Argentina.

Answer Key Chapter Nine

Case Study Exercise:

(1) The dollar difference in median income for the years 2003 and 2004 is a $94.00 decrease.

(2) Married couple families report the greatest amount of median income in 2004.

(3) A female householder living alone reports the least amount of median income in 2004.

(4) There is a $44367.00 amount difference between a female householder living alone and a married couple family.

Group Quiz Exercise:

1. B
2. C
3. B
4. A
5. D
6. A
7. A
8. B

Outside Activity:

(3) According to research, Washington state has the highest paying minimum wage rate.

(4) According to research, Kansas State has the lowest paying minimum wage rate.

Answer Key Chapter Ten

Map the Stats Exercise:

(1)Colombia, Brazil, Argentina, Australia, and Sweden appear to be most open to the idea of accepting other racial categories as neighbors. Mexico, Venezuela, Nigeria, India, and China seem to be the least welcoming.

(2) Competition over resources seems to historically come before racism.

Case Study Exercise:

(1) Holocaust revisionism is the belief that <u>the Holocaust</u> did not occur as it is described by mainstream history.

(2) Over 5 million Jews were systematically killed by the Nazis and their allies.

(3) In addition, most Holocaust denial implies, or openly states, that the current mainstream understanding of the Holocaust is the result of a deliberate Jewish <u>conspiracy</u> created to advance the interest of Jews at the expense of other peoples.

(4) Holocaust denial is also illegal in a number of <u>European</u> countries.

Group Quiz Exercise:

1. B
2. C
3. B
4. D
5. C
6. D
7. B
8. A
9. C
10. A

Answer Key Chapter Eleven

Map the Stats Exercise:

(1) Russia, Asia, Thailand, Cambodia, Botswana, Angola, Mali, Norway, Sweden, and Finland have the highest participation among women in the labor force.

(2) Ecuador, Peru, Argentina, Algeria, Libya, Egypt, Sudan, Saudi Arabia, Iran, Afghan, and India have the lowest participation among women. The United States fits in at 44-45%.

Case Study Exercise:

(1) Rene Redwood was Special Assistant to the Secretary of Labor, Robert Reich (during the Clinton administration). She was executive director for the Glass Ceiling Commission and Greenberg-Lake.

(2) We do not yet live in a color blind or gender blind society. Sexism, racism, and xenophobia live side-by-side with unemployment, underemployment and poverty; they feed on one another and perpetuate a cycle of unfulfilled aspirations among women and people of color.

(3) Of the 5 percent of these managers who are women, only 5 percent are minority women.

(4) Male non-Hispanic whites make up the greatest percentage of the total workforce. Male and female American Indians make up the least percentage of the total workforce.

Group Quiz Exercise:

1. A
2. C
3. C
4. A
5. A
6. B
7. D
8. A

Answer Key Chapter Twelve

Map the Stats Exercise:

(1) According to the map, the majority of nursing home citizens live in the North Central United States.

(2) The least number of nursing home residents live on the West coast. This is do to the high cost of living.

Case Study Exercise:

Case 1

Yes

Case 2

Yes: Symptoms of an abuser include:

    Verbally assaulting, threatening or insulting the older person

    Concerned only with the older person's financial situation and not his or her health or well-being

    Problems with alcohol or drug abuse

    Not allowing the older person to speak for him- or herself

    Blaming the older person

    Attitudes of indifference or anger toward the older person

    Socially isolating the older person from others

Group Quiz Exercise:

1. B
2. B - Gerontology
3. B – They are both important
4. D
5. D
6. A
7. A
8. D

Answer Key Chapter Thirteen

Case Study Exercise:

1) What projection regarding smoking deaths in China was recently made by the American Medical Association?
Another study in the Journal of the American Medical Association stated that smoking-related illnesses could eventually kill 150 million current smokers in China.

2) Why is China the target for large tobacco companies?
China's huge market has made it the prime target of the multinational tobacco companies. Historically, China's cigarette market has been highly protected, with foreign multinationals barred from operating in the country.

3) Describe what has happened with the large cigarette companies recently.

- Philip Morris recently entered into a number of joint ventures with CNTC to grow tobacco as part of an agreement to produce and sell Marlboro cigarettes in both domestic and foreign markets.9
- RJ Reynolds has built a cigarette factory as part of a joint venture with CNTC to produce 2.5 billion Camels, Winstons and Golden Bridges (a local brand) annually.10
- British American Tobacco is involved in a project to increase leaf production with seeds that the company has developed in other countries.
- Other foreign companies have been involved in a number of initiatives, ranging from the introduction of new tobacco seeds, to the importation of high speed cigarette making equipment, to the construction of tobacco processing factories.

4) Describe how smuggling beneficial to large cigarette manufacturers.
Smuggling is also helping foreign cigarette manufacturers enter the Chinese market. Many observers believe that the multinational companies are involved in smuggling as a way for them to develop brand loyalty prior to the full opening of the Chinese market. Smuggling also encourages more people, especially young people, to smoke, since smuggled cigarettes are cheaper, often selling at one third the cost of legal cigarettes.

Group Quiz Exercise:

1. B
2. C
3. B
4. D
5. D
6. A
7. A
8. D

Answer Key Chapter Fourteen

Map the Stats Exercise:

(1) Washington, Oregon, Nevada, Arizona, Colorado, Kansas, Maryland, New York, Vermont, and Connecticut have the highest percentage of women in state legislatures.

(2) South Dakota, Oklahoma, Arkansas, Louisiana, Mississippi, Alabama, South Carolina, Kentucky, Pennsylvania, and New Jersey have the lowest percentage of women in state legislatures.

Case Study Exercise:

1) It was based on analysis of more than 1.4 million contributions of $200 or more given during the 2001-2002 election cycle.

2) The study also found that Democrats were far more reliant than Republicans on deep-pocketed givers in the 2002 election cycle.

3) Looking at giving by gender, the Center found that <u>women who listed an employer</u> or income-earning profession in their contributions during the 2002 election cycle gave 61 percent of their money to Democrats and 39 percent to Republicans.

4) Women who identified themselves as "homemakers" or listed some other non-income-earning occupation preferred Republicans over Democrats by 55 percent to 45 percent.

5) Among very wealthy donors, Democrats reigned supreme. Contributors of $1 million or more gave 92 percent of their money to Democrats.

6) The findings illustrate the Republicans strong advantage over Democrats in the current system, which caps total contributions to candidates.

Group Quiz Exercise:

1. C
2. A
3. B
4. D
5. B
6. A
7. B
8. B

In-Class Activity:

Ans: Watermelon

Answer Key Chapter Fifteen

Case Study Exercise:

Is there anything Blaire's teacher can do to help her?

- **Keep reaching out to the parents.** This teacher was very wise to call Margie's mother in, instead of waiting to see if things would get better on their own. Teachers may hesitate to call parents at a moment like this, not wanting to be intrusive, although a parent-teacher conference frequently turns out to be as meaningful as it was in Margie's case. It may lead to increased rapport, allowing you to move to the next step: recommending counseling.
- **Suggest professional guidance.** Mention to the parent that many people going through divorce have found counseling very helpful. In some school districts, there are regular group meetings for children of divorce. The groups are led by a mental-health professional or a teacher with special training, and children have a chance to discover that their situations are not unique.
- **Provide support for the child.** The support you give a child like Margie on a day-to-day basis is also crucial. A special word of greeting or a warm chat about anything at all is likely to help her feel more at ease.
- **Give it time.** Children in Margie's situation will probably regain their zest for life; just be patient. The first year is generally the hardest for divorcing families.

Group Quiz Exercise:

1. B – extended family
2. B – family of procreation
3. A
4. B - Endogamy
5. A
6. A
7. B - older
8. C
9. D
10. B

Answer Key Chapter Sixteen

Case Study Exercise:

1) In this unusual project, who were the first experimental subjects used in the study?
Deb Roy at MIT's Media Lab, US, devised the unusual project and even volunteered his own family as its guinea pigs.

2) What is this scientist hoping to better understand?
The early development of language.

3) Why is the development of language debated among scientists?
Most psycholinguists agree that simply listening to speech is not enough for a child to piece together the basic rules of a language. Yet they still argue about the importance of specific "language genes" and other non-verbal environmental stimuli.

4) What is "speechome?" Where is the footage viewed?
To provide a more complete picture Roy has developed a surveillance network at his own home in a project dubbed "speechome". Footage recorded by the cameras is automatically transmitted to MIT for analysis.

5) If successful, what might this experiment help in the future?
If successful, Roy says the project could lead to better strategies for diagnosing and treating language disorders. It could even spawn computer programs that can learn to how to speak for themselves.

Group Quiz Exercise:

1. C
2. C
3. A
4. B
5. C
6. C
7. B
8. B

Answer Key Chapter Seventeen

Case Study Exercise:

1) Compare the treatment of Jews in America over time, from post WWII to the present.
For a couple of post-World War II decades, secular Jews were American insiders' favorite outsiders. The increased cultural focus on other minority groups, Christian and otherwise, and the depleted number of secular Jews caused by the full immersion via assimilation by some members of the group into the heart of non-Jewish America. Secular American Jews have less and less of a cultural lectern from which to brandish their particular brand of Americanism.

2) What kind of alienation is the author speaking of, and why?
This alienation is the potential birthright of any American Jew, but the prerogative has been most often publicly exercised by the least religiously observant. The presumption here is that shared tradition, ritual and faith provide observant Jews with a fully accepting community. In short, they have their own world and they are so far from the American mainstream as to not experience the buried but real misgivings about truly belonging experienced by some of their more secular brethren. Of course, at various stages of, or for all of their lives, secular Jews don't feel fully at home in traditional Judaism, either.

3) What is a self-identifying trait for American Jews? Explain.
Jewishness is a crucial self-identifying trait for American Jews. It is in fact that very Jewishness, no matter how idiosyncratically defined, that provides the impetus for the nagging, not always acknowledged feeling of being somehow a half step from receiving full membership in the American melting pot.

4) What role does perception play in this sense of alienation among Jewish Americans?
This feeling of doubt is, at least in my case, not based on any external evidence. It is about perception. More specifically, it is about how you presume other Americans with whom you never had and never will have a conversation about those perceptions perceive you.

Group Quiz Exercise:

1. A
2. C
3. C
4. C
5. D
6. D
7. B
8. C

Answer Key Chapter Eighteen

Map the Stats Exercise:

(1) California, Texas, Illinois, Pennsylvania, New York, Massachusetts, Maryland, New Jersey, Georgia, and Florida have the highest number of AIDS cases.

(2) Alaska, Montana, Idaho, Wyoming, North Dakota, South Dakota, West Virginia, Vermont, New Hampshire, and Maine have the lowest number of cases.

Case Study Exercise:

1) Who funds Canada's health care system? How?
Canada's health care system is a group of socialized health insurance plans that provides coverage to all Canadian citizens. It is publicly funded and administered on a provincial or territorial basis, within guidelines set by the federal government.

2) What kind of care is provided under this health care plan?
Under the health care system, individual citizens are provided preventative care and medical treatments from primary care physicians as well as access to hospitals, dental surgery and additional medical services.

3) Why is Canada's health care system the center of such political debate?
Canada's health care system is the subject of much political controversy and debate in the country. Some question the efficiencies of the current system to deliver treatments in a timely fashion, and advocate adopting a private system similar to the United States. Conversely, there are worries that privatization would lead to inequalities in the health system with only the wealthy being able to afford certain treatments.

4) What is the life expectancy and infant mortality rate in Canada? Why is this important?
Regardless of the political debate, Canada does boast one of the highest life expectancies (about 80 years) and lowest infant morality rates of industrialized countries, which many attribute to Canada's health care system.

Group Quiz Exercise:

1. A
2. A
3. D
4. C
5. A
6. B
7. B
8. C

Answer Key Chapter Nineteen

Case Study Exercise:

1) What is the Bowery and where is it located?
An area Mecca for artists and squatters in Manhattan.

2) What change has occurred on the "lower east side?"
Once a scrappy -- and scary -- Mecca for artists and squatters, the famed Lower East Side street is now "a new millionaire's row."

3) What is one of the hallmarks of gentrification?
One of the hallmarks of gentrification is a sudden, drastic change of rental markets in an area -- usually from working-class people and housing to wealthy professionals and office or retail space.

4) What is the best cure for gentrification?
The best cure for gentrification is prevention.

Group Quiz Exercise:

1. A
2. A
3. C
4. B
5. C
6. C
7. C
8. D

Answer Key Chapter Twenty

Group Quiz Exercise:

1. C
2. D
3. D
4. A
5. A
6. A
7. C
8. B